Natufian and Protoneolithic Bone Tools

The Manufacture and Use of Bone Implements in the Zagros and the Levant

Douglas V. Campana

BAR International Series 494
1989

B.A.R.

5, Centremead, Osney Mead, Oxford OX2 0DQ, England.

BAR -S494, 1989: 'Natufian and Protoneolithic Bone Tools'

ISBN 9780860546320 paperback

ISBN 9781407347837 e-book

DOI https://doi.org/10.30861/9780860546320

A catalogue record for this book is available from the British Library

This book is available at www.barpublishing.com

To Pam, whose constant patience made this possible.

ACKNOWLEDGEMENTS

This project was made possible only through the cooperation and assistance of many scholars. The author especially wishes to thank those excavators who made their data and collections freely available for study. Among them are Ralph S. Solecki and Rose L. Solecki, for the materials from Shanidar Cave and Zawi Chemi Shanidar, respectively; Ofer Bar-Yosef for his material form Hayonim Cave; Tamar Noy for her material from Nahal Oren and Rakefet; and Jean Perrot and François Valla for the material from Eynan / Mallaha.

Several institutions generously provided space and facilities during the course of this research, as well as much welcome assistance. The author wishes to thank also the staff of the Iraqi Museum; the Institute of Archaeology of the Hebrew University, which provided laboratory space over much of the period of field research; the staff of the Department of Prehistory of the Israel Museum; and the staff of the Department of Prehistory of the British Museum. Special thanks are due to J. Zias and M. Dayagi of the Rockefeller Museum for their assistance with the materials from El Wad, Shukbah, and Kebara. The author is also grateful to the staff of the British School of Archaeology in Jerusalem, who provided a home away from home during the research in Israel.

The project was funded by generous grants provided by the National Science Foundation and the Wenner-Gren Foundation for Anthropological Research.

The author owes particular thanks to Ralph S. Solecki and to Ofer Bar-Yosef, for their assistance, criticism, and encouragement over the years. Finally, the author wishes to acknowledge his dept to those scholar, now deceased, whose excavations provided much of the important material for this research; Dorothy Garrod and F. Turville-Petre.

CONTENTS

TABLES

FIGURES

PLATES

INTRODUCTION

This volume presents the results of a research study of the microscopic traces found upon bone implements. These traces are of two kinds. First, when a bone artifact is manufactured the tools used to make it leave distinctive marks upon the bone surface. Second, when a bone tool is used the worked material (or foreign matter present on that material) may leave traces - wear - upon the tool that may indicate the tool's movements during use and in some cases identify the function of that tool. Microscopic surface markings, therefore, can provide information both on how bone tools were made and how they were used. The technology of bone tool manufacture may be characteristic of a culture, and the bone tools themselves may have served a wide variety of subsistence-related and craft activities. The ultimate goal of microtrace analysis is to use these data to form broader inferences about prehistoric technology, economy, and culture. The primary purpose of this study, however, is the development of methods and techniques of microtrace analysis. The application of these techniques to specific archaeological problems is considered here principally as a test of the method, and is therefore of secondary importance.

Such studies were given their initial impetus by S. A. Semenov's now classic work *Prehistoric Technology*, published originally in Russian in 1957 and in English translation in 1964. Although there had been similar attempts before, Semenov's book was the first real effort to put microtrace analysis on a sound methodological footing, and summarized some thirty-five years of experimentation. The fundamental approach of microtrace analysis can be accepted readily on the basis of everyday experience, but there remain many unsolved technical and methodological difficulties. While in recent years microtrace analyses on stone tools have gained some measure of acceptance in archaeology, their proper role in archaeological interpretation has not yet been established. Microtrace analysis is still far from standard practice. Other workers have introduced improved methods for some aspects microtrace analysis on bone since the fieldwork for this project was completed in 1976, but studies of bone implement wear-patterns remain few. The general concepts and procedures discussed in this volume continue to be useful, and it is hoped that they may provide a framework for future research.

The research design of anyone undertaking work in this field must operate at two levels: (1) the specific techniques and methods employed must be tested, and (2) given that the efficacy of the techniques and methods employed can be demonstrated, they must be employed to form inferences based on a specific archeological assemblage. In reality, the most important test of step (1) will be the successful completion of step (2). The researcher must also make explicit the underlying assumptions and theoretical framework upon which the methods and techniques are hung. Whether these assumptions and theory are valid will be tested by the success or failure of the undertaking as a whole.

The above considerations shape the specific problem of this study, which is to test whether: (1) *methods and techniques can be developed which will permit the observation, comparison, and functional interpretation of use-wear and manufacture traces on bone objects*; and, given the above, whether: (2) *the manufacture traces and use-wear on bone implements may be used to assess, characterize, and compare specific archaeological assemblages.* These data may then be used to make inferences about the technological achievements (along with such social and economic effects as may be implied) of the peoples that used the bone tools.

This study represents the application of these principles to the study of two cultures in which bone tools were of major importance, the Natufian culture of the Levant and the Zagros Protoneolithic culture of Iraq. Specifically, the artifacts come from one nearly complete assemblage from the Natufian level of Hayonim Cave, Israel, plus samples from several other important Natufian sites, and the nearly complete assemblages from the Protoneolithic (Zawi Chemian) levels of Shanidar Cave and Zawi Chemi Shanidar, Iraq. These sites are approximately contemporary with the Late Natufian. These bone tool assemblages have great intrinsic interest. In both the Natufian and Zagros Protoneolithic cultures such bone tools appear in large numbers the Near East for the first time. These cultures were entering the earliest stages of settled existence, including intensified food gathering and, in the case of the Zagros Protoneolithic, possible early animal domestication (Perkins 1964). Consequently any additional knowledge of the technological capabilities of these peoples is especially valuable. The wide variety of bone tools, and the high proportions of these tools on the overall assemblages, indicate the importance of bone tools to these cultures. A study of the wear on these bone implements could provide indications for new activities, such as new crafts, as well as evidence for the elaboration and intensification of old ones.

In practical terms, the bone tools examined consisted of the largest readily obtainable samples of bone implements from both these cultures. They include most of the bone tools from these cultures known at the time of this study.

The microtraces on these implements were examined in an effort to ascertain how they were manufactured and the purposes for which they were used. From such observations one may hope to gain insight into the technological repertoire of these peoples, particularly craft activities. A better understanding of their economy, insofar as it was influenced by bone technology, might also be hoped for.

CHAPTER I

Approaches and Methods

Archaeologists' current interest in microtrace analysis has surely stemmed from S. A. Semenov's *Primitive Technology*. But as Semenov himself has stated, his study was by no means the first. Tool-marks and tool-wear are not at all mysterious, nor are they difficult to observe if, as is often the case, they occur on a scale visible to the naked eye. Replicative experimentation was common during the early years of archaeology, and some interpretation of those traces seen upon the artifacts was implicit in these experiments. There have been several reviews of these early experiments (Ascher 1961; Lynch and Lynch 1968; Coles 1973; Graham, Heizer, and Hester 1972; Lewis Johnson 1978). These experiments dealt primarily with lithic implements[1]. While the methods and techniques of archaeology have improved tremendously in recent years, the early workers were quite capable of acute observations. It might well be supposed that in an era when hand craftsmanship and local manufacture were more prevalent than today such markings may have been even more readily interpreted. Such at least is the impression one gains from a reading of Evans'[2] 1872 account of his experimental manufactures of various flaked and ground stone implements.

Most of the earlier archaeological experiments emphasized two major lines of inquiry: artifact replication and experiments with the use of primitive tools. Attempts were made to replicate the gross form of ancient implements and other artifacts using whatever means seemed appropriate to the period to which the artifact belonged. Often these methods were suggested by observations of various peoples who still made use of primitive technologies, such as American Indians. Likewise, ethnographic analogy often suggested possible uses for simple implements of bone and stone. Experiments were conducted to test the efficiency of simple tools for such tasks as the felling of trees, working wood and bone, preparing hides, and many other uses. Evans and other investigators made use of clues provided by marks of manufacture and occasionally wear-marks, but these observations were secondary. The principal aim was to demonstrate that simple tools possessed the potential for performing a variety of jobs.

Semenov regarded tool-marks and wear as his primary sources of evidence. As M. W. Thompson, Semenov's translator, states in his preface, Semenov's approach can be described as analogous to "a sort of trident; the central and main prong is analysis of traces, the two auxiliary prongs are practical

---

[1]Those experiments dealing with bone implements will be discussed in Chapter III.

[2]John Evans has been referred to as the "father of microwear studies," (Tringham et al. 1974: 172).

experiment and ethnographic parallels" (Thompson 1964, x). Semenov also was less interested in the examination of manufacture marks, which had already received considerable attention, than he was in wear-patterns and the determination of tool function.

Semenov's experiments were conducted over a period of many years. Working largely with flint, he found early on that the worked edges of flint implements were marked with striations from which it was possible to deduce the direction in which the tool edge moved during use. Much effort was directed at improving the methods available for detecting and recording these striations. Semenov's work with wear-traces upon bone was much more limited. Following Semenov, in recent years there have been numerous studies of lithic wear patterns. for instance, those of Keely (1977), Odell (1975) Tringham *et al.* (1974) and Vaughn (1985). Very little attention has been given to use-wear upon bone, although there have been several studies of bone tool manufacture, most notably those of Clark and Thompsen (1954), the studies included in the *Premier Colloque International de l'Os dans le Préhistoire*, edited by Camps-Fabrer (1974), studies by D'Errico and Giacobini (1985), D'Errico *et al.* (1984, 1986), and the work of Stordeur (1978, 1979, 1980, 1981).

The place of functional data in the description of artifact assemblages

The form of an implement must obviously be the first consideration in any reasonable assessment of its possible function. Such an implement may be a unique specimen, but it is more likely to be one of numerous examples of similar form within an assemblage. The inference of the function of a single tool contributes relatively little to our knowledge of the lifeways of its maker. A much clearer picture may emerge if we estimate the range of functions represented by all, or at least most, of the tools in an assemblage. Interpretable wear-patterns, however, may appear on only a minority of specimens of a given form. It becomes important to evaluate the implications of wear-patterns discovered on any given implement for other implements of similar form. One must decide whether implements of similar form functioned similarly, or one must confine one's study to those implements with clear-cut wear-patterns. The assumption of similar function for objects of similar form is undoubtedly dangerous, but since functional assessments are already based on inference, the further inference involved may be permissible. It is surely preferable to make such an inference, stating clearly in each case the evidence upon which it is based, than to let it stand as an unsubstantiated implication. The assumption of similar function for implements of similar form becomes necessary if one wishes to construct an inclusive functional classification -- that is, one in which all or nearly all of the implements in a given assemblage are included.

Problems of inference of implement function

The Natufian and Protoneolithic bone tools discussed below may have been used for a variety of purposes. Only rarely can a specific use confidently be assigned to an implement. However, one can often characterize the range of possible uses within fairly narrow limits. Two basic criteria are involved: (1) the formal attributes of the implements must be such that it could have been used to perform the hypothesized job with reasonable efficiency; (2) the wear observed on the implement should support the hypothesized function, or at least not indicate a tool movement or action which would be inappropriate to the task. Judgement on the first criterion, practicality, is made on the basis of common sense (which is, after all, a summary of daily experience) and / or after performing specifically designed experiments. In some cases similar implements will have been observed in use for the proposed function in ethnographic contexts (Gould 1978, Yellen 1977, Kramer 1979).

The second criterion, the observation of wear, demands further discussion. It is important to recognize that a tool's function cannot be inferred directly from the wear appearing on that tool. Two stages of inference are involved, and each has its pitfalls. In the first stage the presence and disposition of chips, scratches, polish, and so on are observed. From these it may be inferred that a certain portion of the tool was its working end, and that it was used with a back-and-forth, side-to-side, or rotary motion, against a hard or yielding material, etc.. Once this has been established, one may proceed to the second level of inference. One may then suggest specific tasks that might have been performed using an implement of a given form, with a specific motion, against a material of a given type. In some cases the range of possibilities is small, and may even be narrowed to one; in other cases the range of tasks of a similar nature is wide, and it will not be possible to pinpoint the use of a given tool within that range.

If the bone implement is complete, or nearly so, and distinct wear is present, this chain of reasoning can be followed with little difficulty, provided that one does not lose sight of its limitations. Such conclusions, however, can only be securely applied to individual specimens. If a function or range of functions can be assigned to a single specimen of a given morphological type, can other specimens of similar form, but lacking interpretable wear, also have that function assigned to them? More generally stated, can a group of artifacts of similar morphology be dealt with as a population when considering questions of function? This additional step represents a third level of inference. Each level of inference must necessarily be less secure than that upon which it was based.

If many similar tools are present several difficulties arise:

1. As the number of specimens increases, the possibility for multiple uses of that tool form also rises. Separating these different functions can be extremely difficult.

2. More specimens mean that the potential range for a differential degree of use increases. Some specimens may become so heavily worn as to be practically "used-up," others may have been lost or broken with virtually no wear. If the choice of individual tools for use was a random phenomenon, then the distribution of the degree of wear should be approximately that represented by a normal curve.

3. As the number of similar tools rises, so does the probability that an individual tool will have been "misused," that is, used for a function other than that for which it was made[3]. One must keep in mind Bordes' (1969: 9) warning that a misused implement is more likely to bear clear signs of wear than one used for its proper purpose. Such an implement must be exceptional, however, or the use represented by its wear-pattern can no longer be considered a misuse. One of the most important reasons for considering tool forms as populations is to avoid being misled by accidental or atypical wear. Conversely, separate specimens may bear different kinds of wear-patterns which are not mutually exclusive, but which may have arisen through similar use of the tool in slightly different circumstances. The discovery of such wear may be taken as corroborative evidence that the hypothesized function is the correct one.

These problems lead us immediately to an additional question. A given wear-pattern can only be expected to occur in part but not the whole of the sample of a tool of a given form. What proportion of the sample must bear that wear-pattern to justify the conclusion that the function inferred from it was the normal one for that tool form? Similarly, what is the highest proportion of occurrences of a given wear-pattern in the sample which may be dismissed as the result or atypical wear? These are extremely difficult questions to answer. Ideally, these limits might be established through experimentation, and experiment can indeed provide useful guidelines. Exactly quantifiable experiments, however, are not possible. Such experiments would depend on precise knowledge of the conditions under which the artifacts being replicated were used. This information, of course, is not available. The best that can be accomplished is to establish broad limits for the occurrence of a given wear-pattern.

Any attempt to treat implement populations contains the implicit assumption that those artifacts can be reliably classed according to form. Such a classification does not necessarily parallel conventional morphological typology. Rather, artifacts which are so shaped as to have

[3]Strictly speaking, if only one artifact of a kind exists, then it cannot be considered to have been misused. Its functional interpretation must be based on its wear-pattern, if any. The purpose for which the implement was made may have been different, and may not be discoverable. If a pair of scissors is used to pry off a bottle-cap, then the proper functional interpretation of that implement, based on its wear, is that it served as a bottle-cap pry; scissors as a class, however, have other uses.

potentially served similar functions must be grouped together. Minor differences in shape are considered only insofar as they could have affected function. Purely stylistic differences, if these can be discerned, are ignored. Such classes must necessarily be broad and flexible. Placement of an artifact within a class is a matter of judgement: could it have served the function hypothesized for the class? If, as analysis proceeds, the original classification proves untenable, the artifact must be reconsidered according to new hypotheses. Because some functions are very similar, classes may overlap. Classes may be divided into subclasses. In addition, during the analytical stage, an individual artifact may be considered as a member of more than one class.

An effort has been made to avoid using designations which have functional implications, or names which through previous usage have taken on functional implications. For example, those objects usually referred to as "gorgets" in the literature are here designated as "small double-points." An attempt has been made to keep the designations as descriptive of form as possible.

These artifacts will be described and analyzed on a group by group basis. Rare or unique specimens will be treated separately. If artifacts of similar form are found in both Natufian and Zagros Protoneolithic sites they will be compared.

Special problems and the methods and techniques used in this project

The literature of wear-pattern studies has included much discussion of the practical techniques involved in the discovery and recording of specific wear-patterns. As nearly all this discussion has centered upon such wear as appears on flint implements, it is to a large extent peripheral to this study. Some workers (Semenov, particularly) have concentrated upon the study of fine-scale striations appearing near flint edges, which necessitates the use of microscopes and cameras of high magnification. Others (notably Tringham) have advocated methods more suitable for the study of large numbers of artifacts and have concentrated on edge-chipping and the use of relatively low magnifications.

The study of wear on bone implements presents its own problems which have had to be dealt with in this study. In addition, a great many practical considerations specific to this project have placed other constraints upon the techniques which could be employed.

The wear appearing on flint implements is usually confined to the area adjacent to the sharp edges. Because of this limitation the use of high magnifications for the study and recording of wear is a practical alternative, although it presents problems when applied to large samples. On bone implements, however, the wear is often spread over very large areas. Therefore, while high powers may be useful in some cases, the complete recording at high magnification of the wear-pattern appearing on an implement becomes a difficult proposition. Consequently, for this study moderate magnifications have been employed, from 50 to 100 diameters for visual inspection and from 5 to 14 diameters (at the negative) for photography. This is believed to be a reasonable compromise given the nature of the wear observed on the implements.

This wear falls into three main categories: (1) smoothing and polish of the bone surface and rounding of edges; (2) chipping and fracture of sharp tips and edges; and (3) scratches, scores, and gouges.

1. Smoothing and surface polish may be considered in two ways: First, its degree and extent may be estimated, and its distribution over the tools surface may be noted. Second, the microscopic topography of the surface may be studied in order to distinguish different types of polish. Both these approaches present practical difficulties requiring more precise instrumentation than that available for this project. Other workers have attained greater precision in these studies since this project was completed (Peltier 1986; Peltier and Plisson, 1985).

The surface polish of interest is largely the result of friction of the implement's surface with a suitable polishing agent, which may be a worked material such as hide or the skin of the hand of the user. In many cases such polishing is probably aided by surface chemical reactions although these have not yet been studied. Whatever the agency, the polish is usually not uniformly distributed over the tools surface but is deeper in areas of heavy friction and shading to nonexistence in other areas.

The distribution of polish, therefore, provides valuable evidence of tool use. It is difficult to record, however. Polish cannot be adequately recorded by ordinary photography, as its presence is only suggested by the presence of glare, which is entirely dependent on lighting. The mapping of polish also presents difficulties as polished areas do not have sharp edges, and its recording is consequently subjective. General observations may be readily made, however, and this has been the approach taken in this study. It is quite probable that all the essential information needed for the reasonable interpretation of function is available from such basic observations, and precise mapping would be useful only in rare cases. This is not to deny that such mapping would be a valuable line for future research.

Similarly, the microscopic topography of polished bone could not be adequately studied in the course of this project for practical reasons. The study of such microtopography has been shown to be useful in the study of flint implements (Keely and Newcomer 1977, Keely 1977), and the approach is also applicable to bone. It is a method, however, that requires the use of very high magnifications (ideally a scanning electron microscope) and as such could not be applied to a comparative study such as this with a moderately large sample. From a purely practical standpoint such equipment was not available under the field conditions of this project, and had it been its use would have been prohibitively time-consuming. Further research is warranted in this area; several more recent studies have made use of the scanning electron microscope for the study of limited samples of bone objects (d'Errico *et al.* 1984; Stordeur *et al.* in press).

On a larger scale, abrasion causes rounding of the sharp edges and point of the bone tool. This rounding is generally easily observable under moderate magnifications, and, if it is confined to a small structure such as a tool's tip or edge, it is readily recorded by photomacrography. The rounding of more extensive areas, such as along the edges of bone shafts or on the projections of an epiphysis, is more difficult to perceive, measure, and record. Like polish, the presence of such rounding has usually been recorded in rather general terms.

2. Tool use may also result in chips or fractures at the working edges or points of bone implements. These chips and fractures are usually easily visible at moderate magnifications, and as they are usually confined to quite limited areas, they are readily recorded by photomacrography.

3. The scratches appearing on bone implements range from those which are clearly visible to the unaided eye to those which may be seen only under high magnification. Such scratches may be distributed widely over the surface of the implement, although fine scratches are generally visible only on those surfaces which have been smoothed or polished by abrasion. On rougher surfaces these scratches are largely hidden by surface irregularities. Patterned scratches which are surely the result of tool-use are visible on many specimens, but these implements have also been subjected to many scuffs and scrapes, both before and after deposition. Scratches are useful primarily because they allow for the inference of the direction of tool movement, but virtually every specimen presenting sufficiently smooth surfaces will be seen to be covered with many randomly-oriented scratches, the results of accidental abrasions. Consequently the observation and photography of very small sections at high magnifications is nearly useless, unless the patterning of scratches on a broader scale can be perceived. In other words, at high magnifications the "noise" of randomly-oriented accidental scratches makes the patterning of use-scratches difficult to observe.

To overcome this difficulty it is necessary to examine the specimen at both high and low magnifications, scanning the specimen for patterned scratches and then recording these scratches in detail by photographic means. Potentially, very fine scratches may only be visible at very high magnifications (300 diameters or more); meaningful scratch-patterns could be seen on a high proportion of the implements at a magnification of about 60 diameters, using a stereomicroscope. These scratches would usually be adequately recorded using stereo-photomacrographic equipment of 5 to 14 diameters at the negative. (Prints of such photographs may have somewhat high final magnifications.) This compromise magnification allows for the recording of most scratches while still covering a sufficient area of the implement so that its form may be perceived and the grouping and orientation of the scratches may be seen. From a purely practical standpoint stereo-photographs at such magnifications may readily be made using portable equipment[4].

It was decided early on in this project that the extensive photographic recording of the wear on the specimens was to be a major aim. The reasons for this decision were as follows: (1) The specimens themselves are widely separated geographically, and although comparison of wear was necessary, this could not be accomplished directly. The recording of wear by other than photographic means can be extremely subjective at best, and it was apparent from the first that perceptions were bound to change with increasing experience. Since the nature of the project did not permit direct re-examination of the specimens at a later date, photography could at least partially remedy this problem. (3) The photographic recording of specimens permits later comparison with experimental implements. This is particularly useful with those experiments carried out after the bulk of the artifactual material had been examined. (4) As will be quite clear in the sections to follow, higher level interpretations become increasingly ambiguous and uncertain. It was desirable to present as much as possible of the original evidence for the reader, to be interpreted as he sees fit. Indeed, this has been the principal object of this study, and higher level inferences are advanced as tentative and subject to reinterpretation.

Specific procedures

Examination and recording of the artifacts proceeded as follows:

After initial grouping into rough categories, the artifacts were painstakingly cleaned. Even museum specimens were found to have surprisingly large quantities of clinging sand and clay. The soil was removed using a soft artist's brush and water. Soil in difficult positions could be dislodged using a small pointed wooden rod. Great care was taken not to injure the bone surface or leave possibly misleading scratches. This work was carried out under a stereo-microscope. A wooden probe, being much softer than the bone, leaves no noticeable marks of itself, but the sand is an abrasive so such deposits must be lifted gently and not rubbed over the bone surface. Unfortunately, one cannot expect such care to have been taken in the original cleaning of the artifact by the excavator; such scratches which may have been left, however, can be expected to be randomly distributed and not related to the working edges and points of the implement. As will be seen in later sections, though, such scuffs and scrapes can lead to difficulties.

After the bone has dried it is then necessary to wash the surface with an organic solvent such as alcohol or acetone before microscopic examination or photography. This must be done because the bone seems to accumulate

---

[4]Stereo-photographs have numerous advantages over conventional photographs for the simple, unambiguous recording of form. See Campana (1977).

an oily surface film which will completely obscure fine scratches unless it is removed. This may be accomplished by gently wiping with a cotton ball soaked in solvent[5].

The position of the patches of polish, edge rounding, and deep scratches and scores visible to the unaided eye are then noted. Photographs of the object are made from as many aspects as necessary to record the basic form of the implement.

The object is next examined at low and then higher magnifications using a stereo-microscope. Manipulation of the light source is important in viewing fine scratches. Many such scratches may be seen if a small bright light source (such as a microscope lamp held at some distance from the object, but sharply focused) is shined at a raking angle over the surface. Very fine scratches which cross highly polished areas may be seen to best advantage by catching the light beam to form a glare. The scratches will be seen as dark lines across the bright patch.

When possible, some specimens were examined using a metallographic microscope at 100 to 300 diameters. This microscope has provision for directing the viewing light downward through the objective. This light will then be reflected back into the microscope by a polished bone surface. Very fine scratches become visible in this way, but high magnifications make interpretation of the scratch grouping difficult as only very small areas can be observed.

After the position of wear-marks has been noted each artifact is photographed in as many aspects as seem necessary to record the observed wear. The worn tips of pointed or spatulate-tipped implements are photographed as a matter of course, so that the degree of rounding and chipping of all tips is recorded.

Photography is accomplished using a specially constructed stereo-photomacrographic apparatus, designed to make a 35 mm stereoscopic pair. This camera is calibrated to maintain a constant magnification, and each roll of film includes an accurate scale, so that measurements may be made from the negatives if desired.

The photography of fine scratches usually requires some manipulation of the light source, which must be small and bright. As in direct observation, fine scratches are most readily photographed against a glare picked up by a polished bone surface. Indeed, such scratches can often be more clearly seen in a photograph than by direct observation, as the exposure at the camera

---

[5]Obviously, any attempt to conserve bone implements using waxy or polymeric binding agents will also leave such a deposit, so that such specimens have little value in such a study. Because the majority of specimens from Mallaha were so treated, they could not be included in this project.

may adjusted to compensate for the brightness of the glare, which can be somewhat blinding in direct observation.

Once the photography was completed a reference collection of 35 mm positive transparencies was prepared, which could be readily handled and viewed for comparative purposes. Approximately 7,000 such transparencies (3,500 stereo pairs) were made during the course of this research.

When recording was completed the data were re-assessed and recoded into a consistent format suitable for computer processing. The measurements and observations gathered on each specimen were cross-checked against the photographic record, in order to compensate for inconsistencies over the period of research. As photographs made early in the research could be compared with those made later, subjective changes in judgement of the visual appearance of wear-patterns could be allowed for. Certain measurements (tip diameters, for instance) were also made at this stage. Generally, the photographs confirmed the visual observations. That is, scratches noted visually were usually also visible in the photographs, and the degree of relative rounding and polish appearing on the photograph was usually similar to that observed by eye. Subjectivity in photography is impossible to eliminate, but one is far more likely to fail to record existent wear-patterns than to generate spurious patterns on a photograph. If distinctive were-patterns were observed visually but had not been recorded photographically, the more conservative course was chosen and the object under consideration was not judged sufficiently worn. In other words, the appearance of wear-patterns in the photographic record was the final criterion used in compiling the data used in the statistical analyses.

The degrees of chipping and rounding of tips and edges were also rechecked against the photographs and ranked on scales running from none to heavy. Actual measurements of such effects could not be made, but frequent cross-checks between photographs ensured that the assignment to the various levels and categories was reasonably consistent over the sample.

Similar recording methods have been used in conjunction with the experimental manufacture and use of bone implements. Photographs of these experimental specimens are presented in this volume for comparison with the artifacts. Descriptions of the individual experiments may be found in the appropriate contexts.

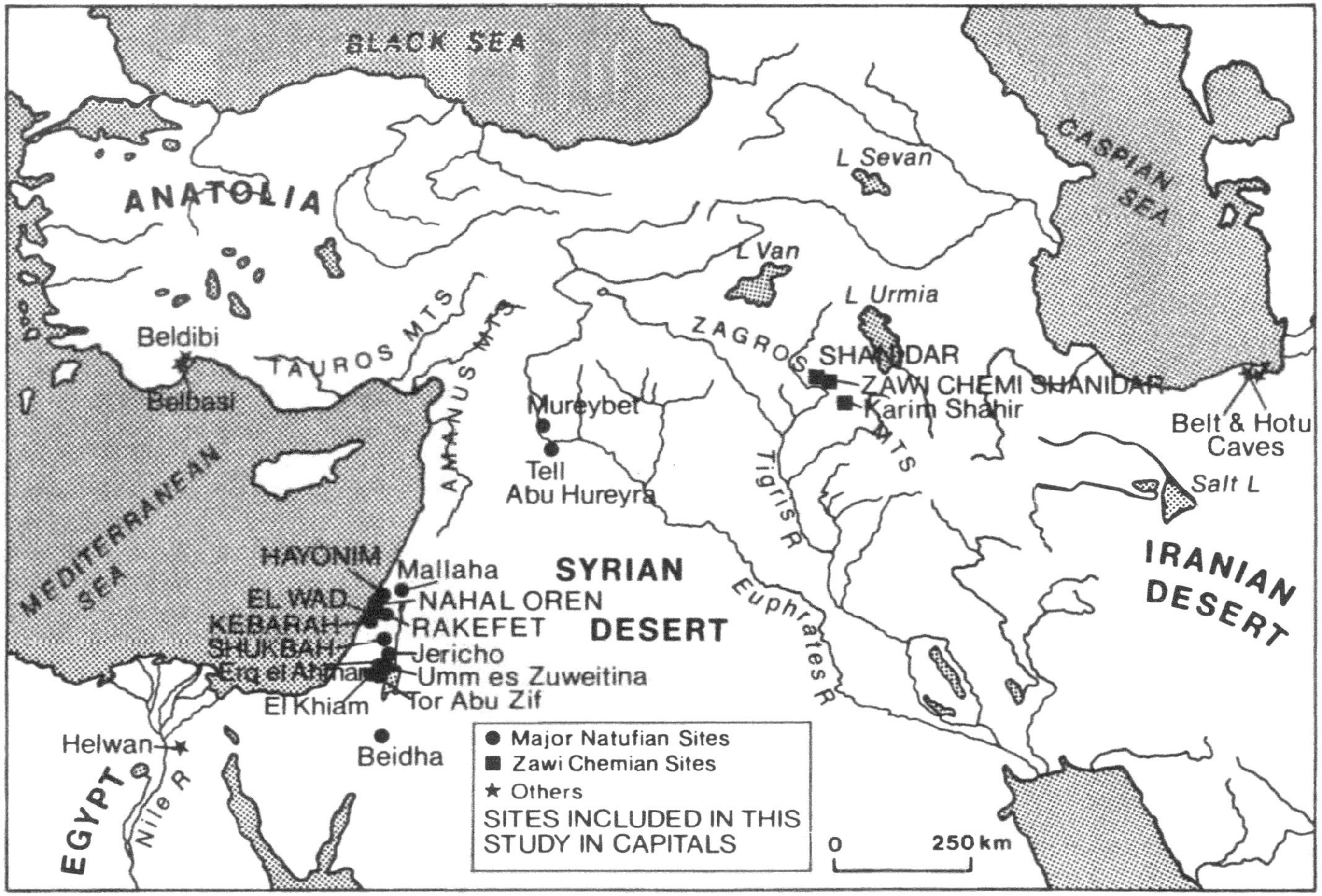

Figure 1: Map of Sites Included in This Study

CHAPTER II

The Specific Archaeological Problem

We turn now to those archaeological problems dealt with in this study. The samples of bone implements chosen for analysis have been drawn from the assemblages of the Natufian and Zagros Protoneolithic cultures of the Near East. These cultures, particularly the Natufian, are well known and need only summary description. This is not to say that these cultures do not present many difficulties of interpretation, and it is hoped that the study of the bone implements may help to illuminate some of the darker areas.

The Natufian culture is the latest of the epipalaeolithic cultures centering primarily in Palestine, with related sites stretching from southern Jordan to Syria. The majority of Natufian sites (including all those treated in this study) are located in the hilly uplands of the Galilee and Judean hills. The hilly upland regions are characterized by a Mediterranean climate, with dry summers and mild, moist winters. Annual rainfall ranges from approximately 400 to 800 mm (Elbashan 1966).

Many Natufian sites are located within the oak-pistachio forest belt which extends over the Palestinian uplands and is continued in a broad arc through the highlands of Lebanon and Syria and the Taurus and Zagros Mountains of Anatolia, Iraq, and Iran. Wild cereals grow co-extensively with this forest belt. In the Palestinian area these cereals are largely emmer wheat and two-row barley (Zohary 1969). It is probable that these cereals and possibly acorns served as major food resources for the Natufian people. Some peripheral Natufian sites extend into the desert regions bordering on the mountain belt.

The most common game animal of the Palestinian Natufian sites was the gazelle (Henry 1975). Red deer, wild pigs, goats, fallow deer, and wild cattle also appear in small numbers, as do various small mammals, birds, and fish. The Natufians evidently took a diversified range of game (Bar-Yosef 1976, Bar-Yosef and Tchernov 1967, Garrod 1932, Henry 1985). Fishing was also practiced -- fish bones have been recovered (in small numbers) from several sites including Hayonim, Mallaha, and Kebara. How important fishing was to the Natufian diet remains an open question.

Chronologically the Natufian follows the Geometric Kebaran A, spanning a period from approximately 10,800 B.C. to 8500 B.C.(Henry 1985). Recent evidence indicates that the Natufian developed without important discontinuity from the earlier Kebaran culture, and also suggests that the Pre-Pottery Neolithic developed from it, similarly without great discontinuity (Bar-Yosef 1970, 1975). Most Natufian sites were abandoned at the close of the period, but some, notably Jericho and Nahal Oren, show some continuity of settlement into the Pre-Pottery Neolithic A, which exhibits many similarities in material culture with the Natufian. Great changes in material culture occurred, however, during the Natufian period. It would perhaps be more fair to characterize these change as quantitative rather than qualitative, as many innovations appear to have arisen earlier in the Kebaran period (Stekelis and Bar-Yosef 1965, Bar-Yosef 1970).

Natufian sites are divided between cave sites and open settlements. Although the interiors of cave were used primarily as places of interment, several cave sites, notably Mugharet el Wad, Nahal Oren, and Hayonim, had extensive occupations surrounding the cave entrance. In the case of el Wad Terrace, this consists of a pavement associated with numerous interments (Garrod and Bate 1937). At Nahal Oren were found many circular stone structures with central hearths which must have been the foundations for round huts. They may originally have had a reed superstructure. Many burials are found between these structures (Stekelis and Yizraeli 1963, Noy *et al.* 1973).

Perhaps the most important Natufian settlement excavated to date is the open site of Mallaha / Eynan (Perrot 1966), near Lake Hula. This extensive site can be fairly considered a village. It is made up of many stone structures with central hearths similar to those at Nahal Oren. The numerous overlapping building levels testify to a lengthy occupation. Some of the structures had floors of lime plaster and numerous plaster-lined storage pits were found. These, unfortunately, have been empty. Many interments were found among the houses.

Size and permanence of settlement are two features which distinguish the Natufian from the preceding Kebaran culture. Individual huts are known from the Kebaran period[6] but generally Kebaran sites are small, from 150 to 400 square meters (Bar-Yosef 1975). In addition, Natufian sites are more widespread than those of the preceding Kebaran. This shift evidently represents both an increasing population and a change in the patterns of subsistence and settlement. Where the Kebarans seem primarily to have been transitory hunter-gatherers, the Natufians are clearly settling into permanent villages or at least into regularly re-occupied seasonal camps.

Natufian sites often contain a large number of burials; about two hundred inhumations are known. This is in marked contrast with the Kebaran, from which only one burial, that of a woman at Ein Gev I, has been recovered (Arensburg and Bar-Yosef 1973). Few grave goods were included in Natufian burials, but some individuals were buried wearing elaborate necklaces and headdresses made from bone and dentalium shell.

The most characteristic Natufian lithics are "lunate" microliths (often with bifacial retouch) and sickle-blades showing silica gloss. Some large tools,

---

[6]Ein Gev I, a hut 5-7 meters in diameter, radiocarbon dated 13,750 B.C. +/- 415 years (GrN-5576) (Henry and Servello 1974), is located on the east shore of the Sea of Galilee (Stekelis and Bar-Yosef 1965, Bar-Yosef 1970).

such as heavy-duty scrapers and picks, were also in use (Henry 1973a, Bar-Yosef 1970). Large stone objects include pecked limestone and basalt mortars and vessels, sometimes quite elaborately decorated. Pestles were also made, sometimes shaped like the hoof of an animal, sometimes shaped like a phallus (Garrod 1932).

Several artifact types which may be regarded as typically Natufian have Kebaran prototypes. Two of these are directly related to the changing modes of subsistence: pecked mortars and pestles, and flint blades bearing sickle gloss. Both are found at Ein Gev I, but they are widespread at Natufian sites. It is a common supposition that the availability of wild grain was a major factor leading to the growing sedentism of the Natufians. These artifacts are usually interpreted as related to the gathering and grinding of wild emmer wheat which grew in extensive stands over much of the area occupied by the Natufians. It is not the only possible use, however. Mortars and pestles may have been used to process the acorns and other nuts which were abundant in the oak forests which were co-extensive with the wild wheat; blades with sickle gloss may have been used to cut reeds rather than wheat.

Another striking feature of the Natufian culture is the appearance of decorative art. The Natufians frequently decorated the objects they used, often with geometric patterns, sometimes with quite naturalistic carving. Many bone tools were decorated, either with abstract patterns or with naturalistic renderings of animals (deer and gazelle). The carved deer appearing on bone hafts from Kebara cave and el Wad Terrace are well known. Small statuettes of stone were also occasionally made. These include a statuette of a gazelle, an embracing couple, and crude heads (Perrot 1968).

The very common shell and bone beads found in the burials, often arranged into elaborate necklaces and headdresses, show a developed taste for personal adornment. The Natufian costume may have been quite elaborate. Such finds suggest that the Natufians were developing a range of craft activities. Very little direct evidence exists, however; organic material has not survived. This is unfortunate, as it would be of great interest to gain the broadest possible picture of the earliest known experiment with a settled existence.

Bone implements have probably been in use since earliest times. In the Levantine area they are found in the Upper Palaeolithic and Kebaran periods, but they are very rare and confined to a few types: points and simple pointed implements. In the Natufian period bone implements become common, and they appear in a wide variety of forms. In addition to the pointed implements and large double-pointed implements ("bipoints") of the previous periods, there appear barbed points ("harpoons"), some large implements with points on either end, a number of very small double-pointed bone slips ("gorgets"), bone hafts ("sickle-hafts") sometimes elaborately decorated, bone hooks, flat implements shaped like bodkins, large antler implements with broad smooth surfaces, and others. This proliferation of tool forms may be related on the one hand to the shift in subsistence strategies, requiring new implements for hunting, fishing, and food gathering. On the other hand, some of these tools were surely used in the newly developing crafts which provided the Natufians with improved clothing, shelter, and living conditions. The function of some of these tools is self-evident (or has appeared so), others are quite enigmatic. Previous interpretations have been made entirely on the basis of form and analogy with other archaeological cultures and with ethnographic sources. Further inferences about Natufian economy and culture have been dependant upon these interpretations as evidence. "Gorgets" have been cited as evidence for the importance of fishing; the decorated "sickle-hafts" have been connected with the great importance of the gathering of wild grain (Garrod and Bate 1937, Garrod 1932). Little real interpretation of those implements likely to be connected with crafts has appeared in the literature. (This is probably the result of many of the most interesting objects having been excavated early on, and never fully published).

The Natufian represents a crucial early stage in the development of a settled society, and probably the earliest beginnings of many of the crafts which characterize the following Neolithic. It may also represent a stage of intensified specialization in the gathering of wild grain foreshadowing the beginnings of agriculture. The artifactual evidence for these developments is very limited. Every effort must be made to maximize the information obtained from the available artifacts.

The Zagros Protoneolithic culture is much less well known than the Natufian. Only two probably related major sites have been excavated, Shanidar Cave and Zawi Chemi Shanidar. These sites are near one another, situated in the high Zagros mountains, in the valley of the Greater Zab River. A third contemporaneous site is Karim Shahir (Braidwood and Howe 1960), a small open site near Jarmo, in the Chemchemal plain, Iraqi Kurdistan.

Zawi Chemi Shanidar (R. L. Solecki 1964, 1981) is an open settlement site approximately 215 by 275 meters in size. The lower layer is Protoneolithic. There the remains of rough stone walls, some of which outline a circular structure. There are also large refuse pits cut into the soil. A Protoneolithic layer is found at Shanidar Cave (R. S. Solecki 1964), 4 km away. This layer contains a cemetery with at least 28 burials of the period. A few of these burials contain grave goods, including some bone implements. The hafted flint knife described in this study was found in one of the burials. Shanidar Cave may have served as a burial place for the people of Zawi Chemi Shanidar, or may have been occupied on a seasonal basis.

The geographic extent of the Zagros Protoneolithic culture is not yet established. Radiocarbon age determinations place it as approximately contemporaneous with the Late Natufian[7]. It has much in common with the Natufian, but also bears a resemblance to the successor to the Natufian, the Pre-Pottery Neolithic A. The two cultures do not appear to have been in contact. The Zagros Protoneolithic culture is preceded by the Upper Palaeolithic Zarzian. A Zarzian industry underlies the Protoneolithic at Shanidar Cave, and it is probable that the Protoneolithic developed out of it. Zawi Chemi Shanidar appears to have been occupied for a long time, perhaps a millennium, and changes occurred over the period. The Zagros Protoneolithic has not yet been internally subdivided. The immediate successor to the Protoneolithic is not yet established.

Like the Natufian sites, the Zagros Protoneolithic sites also lie in the oak-pistachio forest. The general vegetation of the region may be characterized as temperate, with an annual rainfall of 600-1000 mm (Van Zeist 1969). Wild grains also grow in the region including barley and wild einkorn wheat. The grains may have been utilized by the Zagros Protoneolithic people, but there is no conclusive evidence for this.

The predominant animals found at the Zagros Protoneolithic sites include red deer, wild goat, and wild sheep (Perkins 1964). Various small animals, as well as snails and tortoise, were also utilized. There are no large nearby bodies of water, and fishing does not appear to have been important.

Like the Natufian, the Zagros Protoneolithic shows evidence of the beginnings of settlement, at least on a repetitive seasonal basis. Similarly, there is evidence of a shift in subsistence strategy. The elevated numbers of young individuals in the upper levels of Zawi Chemi Shanidar suggest that sheep may have been domesticated (Perkins 1964).

The Zagros Protoneolithic people were also manufacturing a wider variety of artifacts than their predecessors, including ground stone implements. Lunates, triangles, backed blades, and other microliths are present, but the Zagros Protoneolithic is not a predominantly microlithic industry. There are notched flakes and borers, but no sickle-blades. Side scrapers are present but burins are rare. There are numerous large tools such as picks and choppers, as well as oval quartzite flakes. Ground stone artifacts include querns and mullers and pestles. Also present are grooved steatite objects, possibly shaft-straighteners. There are a number of celts with ground and polished edges.

Numerous bone objects were made, both decorative pieces and implements. Some of the implements were also decorated. Like the Natufians, the Zagros Protoneolithic people showed an interest in personal adornment, as some of the burials included necklaces of stone beads, some

---

[7]Radiocarbon determinations:
Shanidar Cave 8650 B.C. +/- 300 (W-667)
Zawi Chemi Shanidar 8920 B.C. +/- 300 (W-681)
(Henry and Servello 1973).

of which were very finely made. In addition to the bone points and pointed implements, spatulate implements, and other tools discussed in this study, the Zagros Protoneolithic people made a variety of bone beads and pendants. Many of the bone artifacts are decorated with abstract incised designs. There is no naturalistic representation as occurs in the Natufian (R. S. Solecki 1981).

It is clear that many new crafts were developing. The manufacture of basketry is directly evidenced by the recovery of a tiny carbonized fragment from a burial at Shanidar Cave (R. L. Solecki 1964, R. S. Solecki 1964).

Some of the bone implements may have been used in food-gathering activities; others may have served in crafts. Simple forms, such as pointed implements, are much alike in both the Natufian and Zagros Protoneolithic; others are quite different. Some Zagros Protoneolithic tools may have served functions similar to Natufian implements, although their forms differ.

The examination of bone implements from these cultures provides several opportunities. As a test of the methods of microtrace analysis on bone implements it allows for the comparison of assemblages made in very different ways and of substantially different forms. One may also enquire whether or not tools of similar form from both cultures are also similarly worn, and presumably similarly used. That is, the degree of relationship of form and function in the bone tools may be investigated. Conversely, one may search for implements of differing forms serving similar functions.

On the level of archaeological interpretation, inferences of the tool function should aid in the comparison of the craft and food-getting methods practiced over a wide area in the Near East. As much of this area was to see the earliest developments of the Neolithic, it would be of great interest to trace the origins and extent of those methods and craft techniques which may be considered adaptive to a settled life.

The bone implement samples were drawn from the following sites (for a total of 513 implements)[8].

Natufian Sites

1. Hayonim Cave (159 implements)[9], a cave site in northern Israel, was recently excavated by O. Bar-Yosef, the Hebrew University, Israel. The site

---

[8]For convenience of reference the artifacts discussed in this volume have been assigned sequential numbers for each site. The site names themselves have been abbreviated as follows: HAY-Hayonim Cave; KEB-Kebarah Cave; ORN-Nahal Oren; RAK-Rakefet; SHK-Shukbah; SHN-Shanidar Cave; WAD-el Wad Cave; ZCS-Zawi Chemi Shanidar. A concordance of these numbers with the excavators' numbers or museum accession numbers appears in the Appendix.

[9]Many more bone implements were recovered from Hayonim Cave since this study was conducted. Most of these have been examined by the author. As none of the conclusions previously reached have needed substantial revision, the additional objects have not been included in this volume. These objects are discussed in other publications (Campana 1987).

is designated as Early Natufian on the basis of flint typology (Bar-Yosef 1975). Hayonim Cave provided the most complete sample of Natufian bone implements, consisting of all specimens excavated prior to 1977. All complete implements and tip fragments were included. Very small shaft fragments which could not be oriented were not included in the core sample. Some specimens which were originally included in this study have since been re-assigned to the Aurignacian period by the excavator. They have consequently been excluded here. These implements are currently in the care of the excavator at the Hebrew University, Jerusalem, with the exception of a few specimens in the Israel Museum (Bar-Yosef and Goren 1973, Bar-Yosef and Tchernov 1967, 1970).

2. El Wad Cave (30 implements), a cave terrace with both early and late Natufian phases, located in the Mount Carmel range, Israel. This famous site was excavated by D. Garrod in the 1920s. Although this site yielded many bone implements, after excavation these artifacts were apportioned by the excavator to many museums and institutions. It was not economically feasible to track down all these widely separated specimens. The examples remain in the Rockefeller Museum, Jerusalem, and another large sample is in the British Museum. Materials excavated by British scholars during the pre-war period have been released for study to outside scholars (Garrod 1932, 1957; Garrod and Bate 1937).

3. Shukbah Cave (21 implements), a cave site in the Wadi Natuf, after which the culture is named. The site was excavated by Garrod shortly before El Wad. It has been assigned, on the basis of lithic artifacts, to the Late Natufian (Bar-Yosef 1975). Like the El Wad artifacts, these tools were apportioned out, but the majority could be studied in the Rockefeller Museum and the British Museum (Garrod 1932, 1957).

4. Kebarah Cave (89 implements), an important cave site not far distant from El Wad. The site was excavated in 1931 by F. Turville-Petre, but unfortunately never completely published. The entire Natufian level of this site was removed by the excavator. Many of the finest Natufian bone artifacts come from this site -- the exceptionally fine carved bone haft is the best known. Unfortunately, many of the specimens have never been reported in print. The majority of these artifacts are currently in the Rockefeller Museum. Most of the remainder are in the British Museum (Turville-Petre 1932).

5. Nahal Oren Terrace (10 implements), an open settlement site spanning both the early and late Natufian phases, has produced only a very few bone implements, although bone artifacts of other kinds are plentiful. These are currently in the Israel Museum, and were examined with the permission of the excavator, T. Noy (Noy, Higgs, Legge, and Gisis 1973; Stekelis and Yizraeli 1963).

6. Rakefet (3 implements), another early Natufian cave site (Bar-Yosef 1975), has likewise yielded only a few implements. These specimens, also in the Israel Museum, were examined with the permission of the excavator, T. Noy.

7. Eynan/Mallaha, a very important late and final Natufian settlement site in the Hula basin, has yielded a few large number of bone implements, mostly in very fragmentary condition. A sample of approximately 100 of the best specimens was examined with the permission of the excavator, J. Perrot. Unfortunately, these implements had been conserved by consolidation with a plastic binding agent, and were not suitable for wear-pattern analysis. In general these implements do not differ appreciably in form or method of manufacture from those found at other Natufian sites (Perrot 1966, 1968).

Zagros Protoneolithic Sites

1. Zawi Chemi Shanidar (171 implements), a Protoneolithic settlement site in the Zagros Mountains, Iraqi Kurdistan. This site yielded a very large sample of bone implements, from which complete implements and tip fragments were chosen for analysis. Small shaft fragments which could not be oriented were not included. The specimens were examined with the permission of the excavator, Rose L. Solecki. The majority of the artifacts are in the collections at Columbia University, New York and the Iraqi Museum, Baghdad (R. L. Solecki, 1964).

2. Shanidar Cave (30 implements), a deep cave site not far from, an probably associated with, Zawi Chemi Shanidar. A small number of bone implements come from the Protoneolithic layer B, which contains many burials. The sample, which has been examined with the permission of excavator, Ralph S. Solecki, is distributed similarly to the Zawi Chemi Shanidar collection (R. S. Solecki, 1964).

Although every effort has been made to maximize the sample of bone implements included in this study, it is of moderate size (513 specimens). These specimens must be divided into several groups based on tool form and possible function. It was decided, therefore, to treat all the similar specimens from Natufian sites together as a group, and likewise to group the specimens from Zagros Protoneolithic sites. Although internal comparisons among the sites of these cultures, made on both geographical and chronological grounds, would be of great interest, such comparisons have been limited here. There are two reasons for this choice: (1) The bone tool assemblages from only three sites, Hayonim Cave, Shanidar Cave, and Zawi Chemi Shanidar, approach being complete as excavated. The samples from other sites are from museum collections from which some fragmentary specimens are missing. These circumstances have been mentioned above. In addition, particularly among the Natufian sites, the standards of recovery and recording have varied greatly over the nearly sixty years since the first sites were excavated. Consequently any quantitative comparisons among these sites would surely be skewed and highly suspect. (2) Both in the

Natufian and Zagros Protoneolithic cases the bulk of the material comes from a single site (Hayonim Cave and Zawi Chemi Shanidar, respectively). The samples from other sites are markedly smaller. Were comparisons to be attempted among these sites only a few implements of a given form would be available from most sites, generally too small a number to be statistically meaningful.

Therefore, although the numbers of the various tool forms from each site are presented here, few statistical comparisons within these cultures have been attempted. No firm conclusions, which could only be based on such comparisons, can be offered.

Both the Natufian and the Zagros Protoneolithic sites surely represent a span of time during which some degree of cultural change occurred. These changes have led to the subdivision of the Natufian into early and late phases (Bar-Yosef 1975; Bar-Yosef and Valla 1979). However, for the purposes of this study these cultures were considered to be sufficiently homogeneous to be treated as units. The Natufian and Zagros Protoneolithic however are distinct geographically and to some degree temporally separated cultures, with numerous typological and technological differences. The samples from these two cultures clearly could not be pooled; on the contrary, as discussed above, they provide useful opportunities for comparison.

The data-gathering phase of this project was funded by grants from the National Science Foundation (Grant No. BNS 75-17913 A01) and the Wenner-Gren Foundation for Anthropological Research (Grant No. 3107). The recording methods and techniques were worked out at Columbia University in the fall of 1975, making use of the portions of the Shanidar Cave and Zawi Chemi Shanidar collections available there. Portable photographic equipment was constructed, as most such special apparatus would not be available in the field.

The field gathering of data was conducted over a period of seven months from mid-June 1976 through mid-January 1977. Facilities for research in the field were kindly provided by the Iraqi Museum, Baghdad, and the Israel and Rockefeller Museums, Jerusalem, and particularly by the Institute of Archaeology, the Hebrew University, Jerusalem, where most of the work was conducted. Similar facilities were provided by the British Museum.

CHAPTER III

The Manufacture of Bone Implements

The Structure and Mechanical Properties of Bone

Bone can generally be divided into four types: long bones (such as the tibia, femur, and metapodials), short bones (such as the carpals and tarsals), flat bones (such as the bones of the skull and ribs), and irregular bones (such as the vertebrae). The great majority of bone implements are fashioned from the shaft portions of ungulate long bones. Long bones consist of a cylindrical shaft, or diaphysis, with broadened articular ends, or epiphyses. The substance of bones is of two kinds. Compact bone, which constitutes the surface of the bone, forms a thick-walled cylinder at the central portion of the diaphysis and is reduces to a thin shell toward the epiphyses. Spongy or cancellous bone, which is made up of interwoven spicules or plates filling the articular ends of the bone, serves as a system of mechanical supports to resist those stresses to which the bone is subjected. The interstices of the spongy bone are filled with marrow.

Bone implements are generally made from the thickened compact bone at the center of the diaphysis. Structurally, this tissue is made up mostly of thin layers of calcitic bone matrix, lamellae, which are concentrically deposited surrounding a central haversian or nutrient canal. These are the osteons or haversian system. The axis of the osteon is oriented parallel to the average compressive stress on the bone. These structures, therefore, are parallel to the long axis of the bone shaft near the center of the diaphysis and splay outward at the epiphysis.

The disposition of the osteons accounts for many of the physical characteristics of compact bone. Where they are parallel, as in the shaft, the bone is markedly anisotropic - that is, the bone is much stronger when stressed parallel to the axes of the osteons than when stressed at a right angle to the axes. Toward the extremities of the bone, where the osteons are no longer parallel, the compact bone is more nearly isotropic in character (Pope and Outwater 1974). Less energy is required to propagate cracks longitudinally in a long bone shaft (parallel to the osteons) than transversely to the shaft axis, with the difference being most marked at the center of the shaft (Pope and Outwater 1972). Accordingly, fragments obtained from the shattering of long bones are most elongated in the direction of the "grain" (alignment of the osteons) of the bone. Since elongated fragments are usually the most suitable for the manufacture of implements, long bones with straight shafts, and consequently maximal areas of parallel grain, are usually chosen. Ungulate metapodials and tibiae are most common.

In composition bone is made up of an organic fraction (collagen), a mineral fraction (principally hydroxyapatite), and water. Bone behaves mechanically as a two-phase composite material (Katz 1971). Within limits its mechanical characteristics can be described on the basis of a material with given characteristics (hydroxyapatite) embedded in another material (collagen) with different characteristics. The resultant mechanical behavior of the bone is not a simple total of the behaviors of its constituents but results from their interactions. A rough analogy would be the behavior of an artificial material such as fiberglass, which derives its strength, flexibility, and breakage-resistance from the interaction of the glass fibers with the imbedding resin matrix. The removal of the organic material from bone leaves a mineral residue which although quite hard is very friable. Conversely, the removal of the mineral matter leaves the collagen behind as a soft, flexible substance. The two substances in combination yield a material of considerable hardness and strength. Compact bone has a compressive strength of about 20,000 p.s.i. and a tensile strength of about 15,000 p.s.i. (Sisson and Grossman 1975). The surface hardness of bone varies greatly depending on whether it is wet or has been allowed to dry. Human compact bone has a Rockwell hardness number of about 35 when dry and only about 7 when wet (Evans 1973), and similar differences can be expected in other bone. This difference can be of considerable consequence in the working of bone.

Experimental and Ethnographic Evidence for Bone Tool Manufacture

There are three main lines of evidence for the methods used in bone tool manufacture. Some insight into the methods of bone working may be gained from an examination of the ethnographic literature. Such references are rare however, for even in those cases in which bone tools continued to be made long enough after contact for the ethnographer to record, they were often no longer made by primitive methods. Since steel knives are usually among the first outside objects to be imported into a primitive culture, many reports record their use in the making of bone tools. Bone tools may also be experimentally manufactured by the archaeologist, using methods suggested by the ethnographic literature, common sense, and the third source of data, the examination of the artifacts themselves. Such an examination, taking into account the overall shape and fine-scale tool-marks left in the course of shaping, can yield a great deal of information. It is necessary to experiment, however, to interpret this information.

Ethnographic evidence

Only the most technologically primitive peoples make bone tools by methods that would have been available to the Natufian and Zagros

Protoneolithic cultures. These methods include the use of water and fire for softening and splitting bone; the use of stone blades, scrapers, and burins for shaving and incising; and of the use of abrasive stones and sand for grinding to shape and polishing.

The initial step in manufacture is to obtain a suitably shaped bone fragment. This is usually an elongated piece usable for making into awls, needles, or projectile points. Bone shafts may be shattered by percussion, but it is difficult to obtain narrow fragments reliably in this way. If long bones are strongly heated in a fire, however, shrinkage and splitting will occur, yielding long, thin fragments. This technique was used by the Micmac for making awls (Wallace and Wallace 1955), and the Salish to produce blanks for needles (Ray 1933).

Fragments may also be obtained by incising and cutting the bone, and to expedite this it may be desirable to soften the bone in some way. Wet bone is much softer than dry bone. The presoaking of bone is a common practice. The bone may simply be soaked in water as the Salish did (Cline *et al.* 1938), or soaked and steamed as the Maori did for softening whale bone:

> The great bone was first soaked in water and then steamed for 24 hours in a long pit, covered over as in a food oven to confine the steam. The bone was then easier to work (Best 1924).

In one account a Bushman presoaked his bone fragment by pushing it into a Tsama melon(Thomas 1959).

The fragment may be whittled to shape with a knife, as did the Copper Eskimo(Jenness 1946). Many of those peoples now making bone tools with steel knives formerly used flint blades or scrapers for this purpose. No clear account of the present-day use of flint has been found. The Arande of Central Australia used stone fragments to shape their bone nose-ornaments (Basedow 1925), while the Andaman Islanders used a quartz or glass flake or a shell to sharpen their boar's tusk spokeshaves (Radcliffe-Brown 1922).

The most widespread method is to rub the bone fragment against an abrasive stone. The Ona and Yaghan of Tierra del Fuego used rough sandstone for shaping whalebone, finishing the work with pumice (Gusinde 1931), while the Aleuts used andesitic lava for fine smoothing (Jochelson 1925). The Micmac (Wallace and Wallace 1955), the Salish (Ray 1933), the Ojibwa-Chippewa (Densmore 1929), the New Guinea Highlanders (Aufenanger 1961), and the Andaman Islanders (Radcliffe-Brown, 1922), among others, also used the abrasive method.

Some tools must be perforated. The Maori perforate their bone fish-hooks by means of a flint point mounted on flywheel type of pump drill (Buck 1952). The Kuma of New Guinea perforated pigs' ribs with a rat's tooth twirled in the fingers (Aufenanger 1961).

The final step may be to polish the bone object. In addition to pumice, the Salish used horsetails, *Equisitem nymale*, to polish horn (Cline *et al.* 1938), as did the Aleut (Jochelson 1925). This plant is also known as scouring rushes, and it contains a concentration of fine abrasive silica in its stem.

The Experimental Approach

Several investigators have tried to make bone tools by primitive methods. A few studies of particular interest to this project will be briefly summarized here. The first step is to obtain a suitable fragment. Biberson and Aguirre (1965), interested in Lower Paleolithic tools, experimented with fracturing and attempting to split elephant bone, using wedges and hammerstones. They found it very difficult. A broad study of general application was done by H. Sadek-Kooros (1972). Sadek-Kooros' principal aim was to determine whether the shape of bone fragments broken randomly would be significantly different from the shape of bone pieces broken with the intent of producing a form suitable for implements. Some four hundred sheep metatarsals were shattered in various ways, mostly intended to simulate natural processes. Striking a metatarsal with a hammerstone on the medial or lateral surface, followed by twisting the shaft resulted in a form closely resembling a common archaeological type: a pointed implement formed on a metapodial which retains one of its epiphyses, presumably as a handle.

As many European Upper Paleolithic and Mesolithic implements were made from antler, particular attention has been paid to this material. Stimulated by the numerous antler points and partly-worked antler fragments found at the British Mesolithic site of Star Carr, Clark and Thompson (1954) experimented with the extraction of suitable blanks from red deer antler. They made use of a method of widespread importance, the "groove-and-splinter" technique. In this method two (or more) deep parallel grooves were made longitudinally in an antler, using a flint burin as a cutting implement. The remaining central ridge or splinter would serve as the blank for further working. Its length may be limited and its removal facilitated by making transverse cuts at either end. It is then prized up. A cord may be passed under the splinter and worked back and forth to aid in its removal. This appears to have been done with the Star Carr artifacts.

Other workers have noted the necessity for pre-soaking antler before working. R. Feustel (1973), after having soaked an antler for twenty hours, sectioned it by grooving all round with a quartzite cutter. He noted that the upper layer cut easily but that the inner layers were harder and damaged the cutter. Similar experiments were conducted by A. Billamboz (1977), who noted the extreme difficulty of cutting into antler which had not been pre-soaked. Soaking in boiling water greatly softened the antler and eased working, but the antler quickly regained its original hardness, necessitating repeated treatments.

M. Dauvois (1974) replicated antler harpoons of Magdalenian and Azilian types. Blanks were first prepared from caribou antler, using a groove-and-splinter technique. The antlers were soaked in water during working, and the grooves were cut with flint burins, flakes, and fragments of convenient forms. A cutting edge which was nearly right-angled was found most effective. Once these blanks were prepared, similar tools were used to cut the barbs. Dauvois noted that the pressure needed in cutting and the irregularities of the surface led to tool vibrations. These resulted in markings on the antler surface oriented transversely to the direction of the tool movement. Such chatter-marks appear to be a characteristic effect of working bone with flint.

The work of M. Newcomer (1974a, 1974b) is of particular interest to this study. Newcomer investigated the bone and antler industry from Ksar Akil, Lebanon. These artifacts, primarily points, from the Levantine Upper Paleolithic, were antecedent to the various Natufian forms discussed here. In the course of his study Newcomer made about twenty points from cattle and sheep long bones, antler, ivory, and a hippopotamus tooth. Most of these tools were worked using flint and obsidian burins, scrapers, and blades. Few if any differences in the results or the surface markings left on the bones were observed. Three points were ground to shape using sandstone. Newcomer found it relatively easy to distinguish the ground tools from those made with flints, but he did not believe that finer distinctions were possible as had been suggested by Rigaud (1972). Rigaud had conducted similar experiments using bone and horn, and he claimed to be able to distinguish marks left by burins from those left by end-scrapers and blades.

F. d'Errico, B. Giacobini, and P. F. Puech (1984) have published a set of criteria for distinguishing various methods for working bone. These criteria are substantially similar to those used in this project. The present study, however, was largely conducted between 1976 and 1980, and made use of an independently derived set of criteria. It is important, therefore, to describe these criteria.

A variety of bone implements were fabricated during the course of this project. These experiments have been aimed primarily at providing comparative data regarding the specific surface configurations and fine-scale markings seen in the Natufian and Protoneolithic bone tool samples. The experimental implements were used as aids in distinguishing bone objects manufactured by abrasive techniques from those shaved to shape by a sharp, presumably flint, edge.

Materials used for these experiments consisted of fresh cattle bone obtained from a butcher and cattle bone which had sun-dried for at least a year. The bone was worked with flint tools of various types, made from nodules obtained from the Levant and England. Another series of tools was ground to shape using sandstone of various kinds. These processes leave distinctive markings on the bone surface, which, when compared to the artifacts, permitted the method of manufacture to be identified for most of them.

Surface Markings Left by Flint Implements

Fresh bone, dry bone, and dried bone which had been soaked in water held near the boiling point for twenty-four hours were worked with flint scrapers and flakes. In this preliminary trial large sections of cattle long bones (tibia, humerus, and femur) were used. The fresh bone was obtained from a butcher; the dry bone was obtained from a cow skeleton which had sun-dried for at least a year and was thoroughly desiccated. Some of the dried bone was then soaked in water. A half dozen specimens of each type of bone were tested. The largest specimens were hand-held during working; the smaller pieces were gripped in a hand-held clamp.

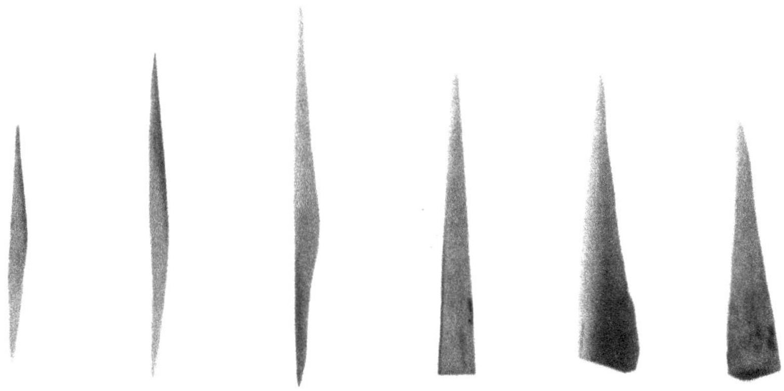

Plate 1: Experimental Bone Tools Made Using Flint Scrapers (1X)

Simple flint implements were made for these experiments, including retouched and unretouched flakes and thick scrapers. At least one specimen of each of the bone types was worked with each of these tool forms. The larger of the flint tools were hand held, while the smaller flakes were hafted in a hand-held clamp.

As expected, dry bone proved the most difficult to work. Despite an applied pressure of several kilograms, the flint edge would scarcely penetrate the surface of the dry bone, but rather would skim over the surface, leaving only superficial scratches. At the pressure required to produce any effect upon the bone, flint flakes with acute edge angles (approximately 15 to 25 degrees) would break or chip heavily, usually upon the first stroke.

The bone which had been soaked in water was quite soft and easily worked for about one millimeter inward from the surface, but the inner layers were as hard as that bone which had not been soaked. If such bone were to be used in fabricating tools, it would necessitate continuous re-soaking to keep it soft enough to work. Fresh bone appears to be the most favorable material for the making of bone articles. It is soft enough throughout to be worked without extensive pretreatment.

Plate 2: Flint Scrapers Used in Manufacture Experiments (½X)

As Newcomer also observed (1974b), thin flint flakes or blades with sharp, acute edges were rather unsuitable for bone working. Much better results could be achieved using rather large, thick scrapers. Thick scrapers with a strong cutting edge having an edge angle of about 90 degrees cut into the bone surface effectively. Such tools could be used for thirty or forty strokes before their cutting efficiency was seriously reduced by edge wear. Small chips or flakes were removed from the flint edge during the course of use, usually from the tool edge opposite the direction of tool movement. Slight rounding of the tool edge, however, appeared to be the principal cause for the loss of cutting efficiency. The edge of a tool could easily be renewed by retouching. Such a retouched edge leaves a characteristic pattern of striations on the bone surface.

The work could be greatly speeded by frequently flushing the bone with water. This washes away the bone chips and dust and lubricates the surface, and possibly softens the bone slightly. Lubrication provides a much faster, smoother, and more even cutting action and greatly reduces wear on the flint edge. Bone worked in this way only sporadically exhibited prominent chatter marks. Newcomer observed chatter-marks on all his specimens, and he considered it a diagnostic feature for working with a flint edge. He does not seem to have lubricated the work surface, however. He attributed the chatter to the flint edge bouncing over minute irregularities in the bone, but this effect is likely to have a more complex cause. Chatter is a fairly well-understood phenomenon met with in the machining of metals. It is caused by the tool edge being forced too deeply into the workpiece. As a result, instead of a smooth shaving being produced, an improperly shaped chip is formed, the bit is stressed and the bit edge bounces out of the work. Regular repetition of this action leads to chatter. Chatter of this sort is the effect of a combination of excessive tool pressure and a tool edge with too steep an angle of approach to the work. The correct combination is determined by the hardness of the material being worked. A similar effect occurs if there is too much pressure on the tool and insufficient lubrication. This leads to "stiction", in which the tool alternately binds and slips as the tool pressure builds to overcome the coefficient of friction of the surface. A familiar example of this is the vibration which results when a dry finger is run over a clean plate. If the tool angle and pressure are properly regulated and lubrication is provided chatter can be avoided.

Following the initial trial, about thirty bone points were made using flint tools, all from fresh cow bone. Blanks, roughly similar in shape to those produced by a groove-and-splinter technique, were cut from bone shaft sections using a metal saw. These blanks were firmly gripped in a clamp and worked to a point using hand-held flint scrapers. The bone blank was sharpened by pushing the flint scraper unidirectionally toward the point, in a manner similar to sharpening a pencil with a knife. The scraper appeared to cut best when the leading plane of the cutting edge was held at an angle of from 70 to 90 degrees to the surface of the bone. Working was continued until the initial shape of the blank was obliterated and the sharpest point possible was obtained. Working was eased by frequently flushing the bone with water. Dependant upon the character of the blank, each point would take from ten to fifteen minutes to complete.

The characteristics of bone worked in this way may be described as follows:

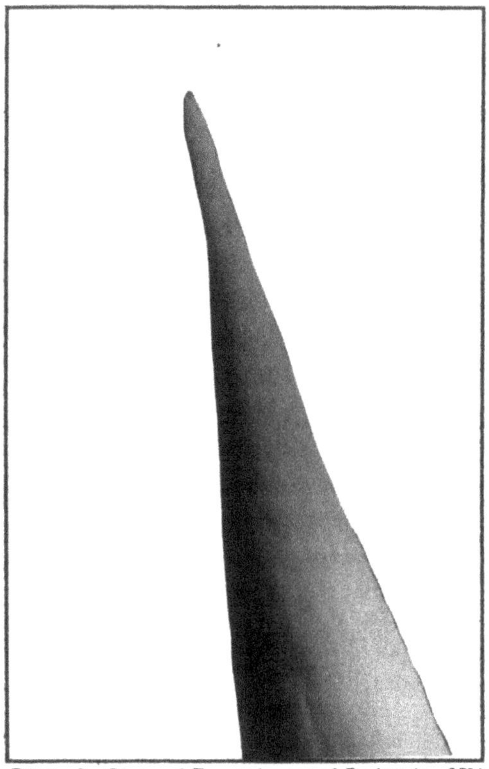

Plate 3: Shaved Experimental Point (c. 3X)

1. **Gross Form**: Because the bone object is worked with a single edge, and since either the tool or the workpiece or both are hand-held, the form of the object is rarely perfectly regular. The flint edge rides up and down on the bone surface, causing small convexities and concavities along the length of the stroke. These are further accentuated on subsequent strokes, leading to a cut with an undulating profile. This slight waviness is irregular rather than vibratory and can be distinguished from chatter. If the implement being made is cylindrical, such as a point, these small undulations rarely coincide as the flint edge is used on subsequent strokes around the bone cylinder. The bone object, therefore, tends to be slightly irregularly shaped, dependant, of course, on the skill of the bone worker. If the bone is cut to shape with a relatively few deep cuts the resultant cylinder will have a rather polygonal cross-section. If the bone surface is hard and worked slowly with multiple strokes, a low tool angle, and light pressure a very regular shape may be achieved. The lack of undulations cannot be taken as evidence that an artifact was not shaped with flint tools.

2. **Fine Markings**: When examined with a hand lens or a stereoscopic microscope the bone surface shows a distinctive pattern of tool-marks. A flint edge is almost never perfectly straight or regular. It may have been intentionally retouched, so that it bears a series of characteristic small flake scars. An unretouched edge is soon damaged in working, so that it quickly

develops numbers of small nicks and scars along its length. When such a tool is pushed down the length of the bone workpiece these scars leave parallel

Plate 4: Surface Markings Left by Flint Scraper (c. 24X)

striations on its surface. If the worked piece is cylindrical or conical in shape several of these striations can be seen running down its length. These will be cut across by other sets of striations left by subsequent strokes. The striations are rarely straight, but rather tend to undulate back-and-forth due to lateral tool movements. Shallow, rapid stokes with light pressure produce fine striations which are nearly straight, while deeper, slower movements produce more pronounced undulations.

When seen in cross-section, the striations left by a flint edge tend to have rather flat or shallow curving bottoms. These areas often have a somewhat glossy appearance. On the objects most commonly encountered this area is quite small. Only the facet of the tool which is in actual contact with the bone can be responsible for these markings. In most cases, therefore, it will be difficult distinguish what specific type of flint tool left the markings.

Abrasive Methods:
Characteristic Surface Markings

Bone can be very rapidly and effectively shaped by rubbing against an abrasive stone. A series a preliminary trials were conducted which were essentially similar to those described above for working with flint tools. The abrasive stone used was a sandstone having well-consolidated sharped-edged grains.

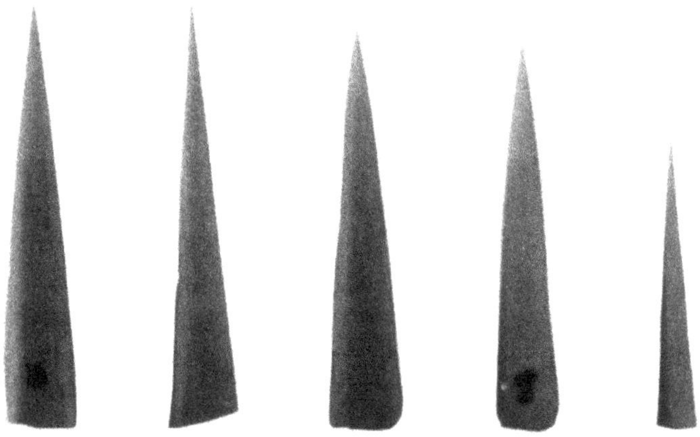

Plate 5: Experimental Bone Tools Made Using Abrasive Technique (1X)

One may distinguish two basic bone grinding approaches. Both these techniques were tested:

a. The bone fragment may be rubbed back-and-forth across the abrasive stone at or nearly perpendicular to the axis of the bone fragment (which is almost always parallel to the axis of the bone shaft). This will be referred to here as cross-grinding. Cross-grinding was found to very effective for rapidly reducing and rounding the sharp edges of the bone fragment, for rough shaping, and for forming flattened facets and spatulate tips.

b. The bone fragment may be rubbed in a direction parallel to its length (usually parallel to its grain). This will be termed axial grinding. This action results in much slower cutting than cross-grinding, but it leaves a smoother surface with even curves. It is especially effective for making sharp tips and for shaping regular, symmetrical objects such as points.

There appeared to be little difference in the effectiveness of grinding techniques related to whether the worked bone was fresh, dry, or water-soaked. It is particularly important, however, to keep both the bone and the sandstone rubber continually flushed with water. Working under running water is best. If this is not done the sandstone quickly becomes clogged with the sticky bone waste and soon ceases to cut. The sand grains at the surface also rapidly break down. If the stone is kept flooded with water these waste products are removed, presenting a continuously fresh grinding surface.

Cross-grinding appears to cut more rapidly than axial grinding because, as the angle of relative movement between bone and grinder is not fixed, slight changes of stroke direction result in the abrading away, with each successive stroke, of the ridges left by the preceding stroke. The angle of tool movement is more strictly fixed in axial grinding and this effect does not occur.

Plate 6: Sandstone Grinder: Note (1) Narrow Groove, Made by Axial Grinding; (2) Saddle-shaped Depression Made by Cross-Grinding (½X)

About thirty experimental bone points and ten bone implements with spatulate tips were ground to shape from fresh cow bone. The sandstone grinder was placed on a table and the hand-held bone blank rubbed back-and-forth against it. Both the bone and sandstone were flushed frequently with water. A bone point with the sharpest possible point could be made in approximately five minutes. Tools with spatulate tips were made from large fragments of cow bone. In some cases the epiphyseal end was left attached to serve as a handle. The method was generally similar to that for the points, except that the tips were cross-ground. Spatulate tips could be made in from five to ten minutes.

The characteristic forms and surface markings of objects made by axial and cross-grinding methods are summarized below:

1. **Gross Form**: Bone objects shaped by cross-grinding usually show clear-cut flats or facets, often with a distinctly angled corner between them. Cross-grinding with small or curved grinders can produce hollows or concave areas, but these will have a cylindrical cross-section. Axial grinding tends to produce smooth, regular curves rather than flats. It is difficult to produce concavities on small bone fragments with axial grinding; straight cylindrical, conical, or slightly convex forms are more typical. Axially ground bone objects tend to be quite symmetrical, with smooth, even curves and surfaces.

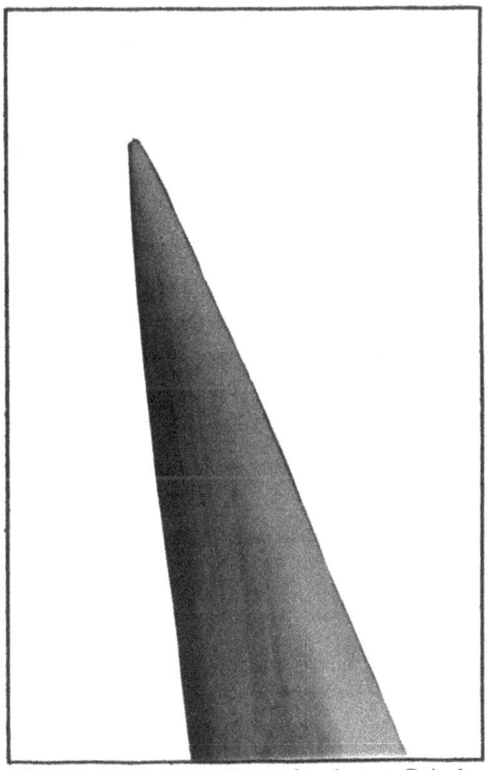

Plate 7: Point Made Using Sandstone Grinder (c. 3X)

2. **Fine Markings**: Cross-grinding tends to leave clear, parallel striations running across the facetted areas. These striations are usually straight, fine, and evenly spaced. They are quite close together, depending on the coarseness of the abrasive surface. The striations tend to have V-shaped bottoms. Axial grinding produces similar striations, but because the bone being worked is usually conical, groups of striations tend to overlap,

leading to a criss-cross pattern. Very fine axial grinding may be difficult to distinguish from the natural grain of the bone surface.

Plate 8: Surface Markings Created by a Sandstone Grinder (c. 24X)

Other techniques

Several other techniques deserve a brief description although they were not the subject of extensive experimentation in this study. The sectioning of bone to make blanks, the groove-and-splinter technique, and other needs may require that grooves be cut through the bone wall. When the cut must be made parallel to the bone axis this may conveniently be done with a flint burin or the broken end of a flake or blade which presents a facet similar to that of a burin. Several attempts were made to cut such a groove into the surface of a cow bone using flint flakes and blades with a burin-like facet of about 90 degrees at one end. The bone shaft was held in one hand and the flake or blade in the other. With the burin-like facet at nearly right-angles to the bone surface the flint edge was pulled longitudinally down the length of the bone shaft. As with shaving, keeping the surface flooded with water aided the cutting action. Grooves cut in this way tended to have rounded or V-shaped bottoms. The shape of the groove only partially reflected the shape of the cutter as many strokes with considerable tool movement and overlap between them were needed to complete the grooves. The inner surface of the groove was usually scored with semi-parallel, partially overlapping striations running along its length. As it was often difficult to start a groove, there would commonly be a number of scars and scratches from false starts running parallel to it. The groove would not be perfectly straight, but curved with the surface of the bone.

Grooves made around the circumference of the bone shaft may be cut with a burin facet or with edge of a flake or blade which may be denticulated and used in the manner of a saw. Several such cuts were attempted making use of a flint flake with an edge which had be denticulated by retouching with a bone point. Such cuts were similar to those described above but tended to be somewhat straighter. When made with a flake or blade the striations sometimes ran tangentially to the circumference of the bone shaft rather than around it.

A bone shaft may also be cut by means of a taut cord which has be charged with a slurry of water and sand. As the cord is drawn back-and-forth the sand abrasive gradually cuts into the bone. This method is quite efficient and can be used top cut through hard stone but may be slow and require patience. Billamboz (1977) discusses its use in the French Neolithic for cutting through antler. Poplin (1974) describes its use by the Predynastic Egyptians to sever a hippopotamus tooth. No experiments with this method were conducted in this study but Poplin experimentally cut through a tooth using this technique. Briefly summarized, such a cut appears quite straight with parallel walls. Some circumferential striations may be visible, caused by the cord being worked around the object rather than cutting straight through from a fixed position. The walls of the cut may show slight undulations caused by the movements of the cord, which does not necessarily cut straight down as does a metal blade. The base and wall of the cut gain a smooth, polished appearance.

Flint borers or suitably shaped flint flakes may be used to make perforations in bone. Several experimental holes were made in this way, using flint flakes with two opposed sharp edges as borers. If the flint implement was hand-held or mounted in a simple haft a hole could be cut with a rotary wrist-action but the job required a good deal of effort and was time-consuming. Depending on size, a single hole could take an hour or more to complete. As the flint penetrates the bone the area of tool edge contacting the hole wall increases, making the work increasingly difficult. Consequently it is often convenient to drill through alternately from either side of the bone. The resultant hole will be left with an hourglass shape in cross-section. The borer will leave striations running around the circumference of the hole. Depending on the care of the worker, these may completely or partially surround the hole. The striations in a hand-cut hole are often at an angle to the axis of the hole. Hand-bored holes tend to be slightly irregular in shape and the walls are rarely parallel.

The drilling can be speeded many-fold if the flint cutter is mounted in a rotary drilling device such as a bow drill or pump drill. Holes made in this way are more commonly cut from one side only. Several experimental holes were cut with a flint borer attached to a rotary brace. They were more truly round and the walls were more truly parallel than holes cut with a hand-held borer. The striations within the hole made true, concentric circles. While the edge of a hand-cut hole was often rounded or irregular due to the free movement of the borer, a hole made with a rotary drill would often have a sharp, angular rim.

A rotary drill may also be used with a reed charged with sand and water. This is a common method for drilling holes in hard stone. The holes made by this method are well-formed, rounded, and smooth. No experiments with this method were performed.

A final note concerns the use of heat-treatment to harden bone implements. This possibility has occasionally been suggested in the archaeological literature. As the discussion of the mechanical characteristics of bone would suggest, moderate heating which will desiccate the bone will indeed harden it considerably. Stronger heating however will result in the breakdown of the bone's organic fraction, with the consequent loss of resilience and tensile strength. Increasing temperature will carbonize the bone and eventually calcine it. Carbonized and calcined bone have little strength, and the surfaces are quite soft. Specimens are occasionally found in this condition but the extreme heating must almost surely have been accidental or the result of a process other than one intended to harden the bone.

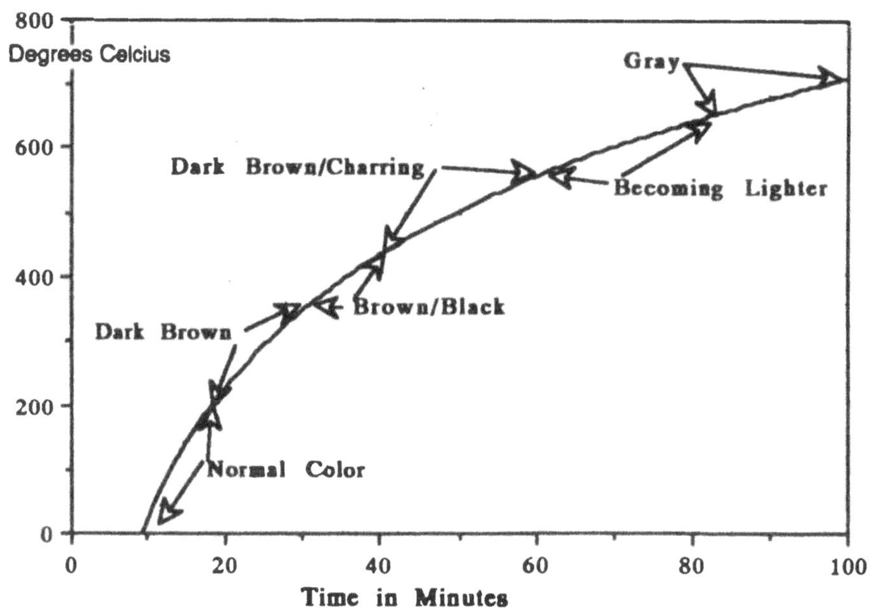

Figure 2: Color Changes of Dry, Compact Bovine Bone with Increasing Temperature Over Time

The color changes of bone with increasing temperature are present in Figure 2. Holding a bone fragment at a given temperature for a prolonged time does not usually result in increased charring or calcination beyond that indicated in the graph.

Natufian and Protoneolithic Manufacture Methods

The methods used for the manufacture of the Natufian and Zagros Protoneolithic implements samples in this study were evaluated according to the above criteria. The bone tool kits of both cultures are rather similar, in that the bulk of the assemblages for both consists of simple pointed implements of similar form. The manufacture techniques, however, proved to be strikingly different. At the most basic level, while Natufian bone tools were mostly shaped using flint implements, the Zagros Protoneolithic tools were worked almost entirely by abrasive techniques.

In terms of efficiency, the differences between shaving and grinding are minor; grinding appears to be somewhat faster and to produce well-shaped objects more easily. Shaving techniques were surely in existence from the earliest times for the working of wooden objects. The innovation of abrasive techniques, when applied to stone artifacts, was the start of the Neolithic, according to the original (Lubbock 1865) definition of that term.

Although there is a great disparity in the predominant techniques used for making implements, abrasive techniques were also used by the Natufians (commonly to produce objects of personal adornment), and likewise flint shaving was sometimes used in the Zagros Protoneolithic. The difference, therefore, was not due to a lack of awareness of alternative methods.

Natufian Manufacturing Techniques

The Natufian bone tool assemblage is characterized predominantly by simple pointed implements and fragments of such implements. Awl-like tools of various sizes, made on longbone shafts, are also common. Less frequent forms include small bone bipoints, barbed points, and bone hafts for flint blades (presumably sickle hafts). A few bone hook-shaped artifacts have also been found. Rare forms include elongated tools with a flattened cross-section that may have been basketry or weaving tools. A unique "shaft-straightener" made from a scapula has been described elsewhere (Campana, 1979).

Most Natufian bone implements are made of fragmentary or highly modified bone. The size and shape of most fragments suggest that the predominant material was the long bones of moderately large ungulates. As those specimens which can be identified to species are largely gazelle, this is also likely to be the source of the fragments. A few tools were made from the long bone shafts of small mammals, bird bone shafts, and the spines of fish vertebrae. Large implements were commonly made from the metapodials of artiodactyls such as gazelle. Such tools include either the proximal or distal epiphysis, which serves as a handle. The largest tools may include the entire epiphysis (usually of the distal metapodial) and the complete shaft, which is broken diagonally to form a pointed end. The break on these implements is similar to the twist-fracturing method described by Sadek-Kooros. Unwanted edges or protrusions were occasionally removed by abrasion against a very coarse stone. The tip end was then brought to a sharp point with a flint edge, rather like sharpening a pencil. Such implements show no indication of previous preparation of the shaft.

Another common form requires that a metapodial (usually of a gazelle) be split lengthwise, leaving one half of the distal epiphysis to serve as a handle at the base of the tool. To do this grooves were cut most of the way through the bone wall, on opposite sides of the shaft. Judging from the traces remaining on the tools, the implement used to make the groove had a burin-like tip. The tip of this implement was guided down the length of the bone by running it in the suture line between the fused halves of the metapodial. Those traces that remain on the bone tools appear near the distal end; it is not clear whether the grooves continued down the entire length of the shaft or were brought together to make a point. The subsequent sharpening of the tip removed this evidence. Unfortunately, no partially worked specimens are available. Cutting a groove with a burin requires a substantial effort, so the grooves were rarely cut all the way through. When they were sufficiently deep the bone was split, presumably by wedging.

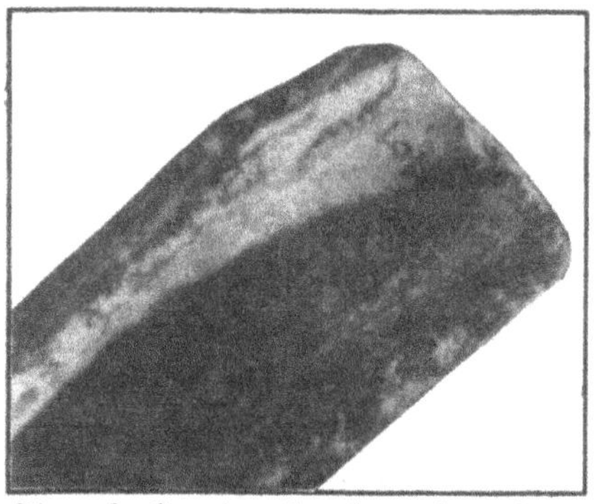

Plate 9: Section Cut from Longbone (HAY 104) by Means of Burin-Grooves and Splitting (c. 2X)

The proximal ends of metapodials were also made into pointed implements. For these tools the bone shaft appears to have been approximately quartered longitudinally, also by burin-grooving. It is possible that these tools could have been made simultaneously with those made from the distal ends, but this cannot be demonstrated conclusively.

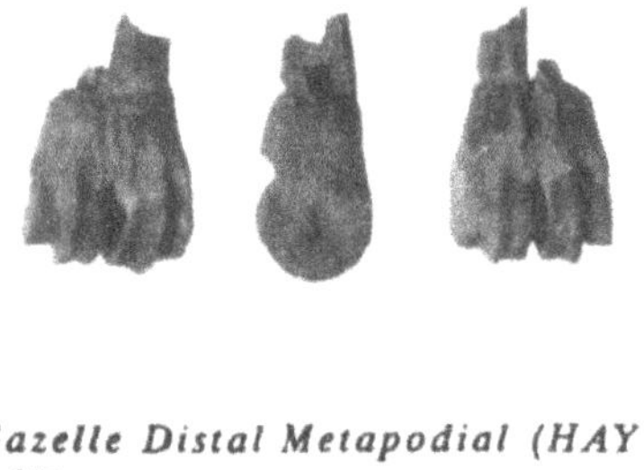

Plate 10: Gazelle Distal Metapodial (HAY 179) left as Waste (c. ½X)

Plate 11: Grooves Used to Remove the Epiphysis and Split the Shaft (c. 2X)

For some tools the distal epiphysis was cut from the shaft before or at the same time as it was split longitudinally. The groove around the shaft appears to have been made with the edge of a flake rather than with a burin. The cut-off epiphyses were left as waste fragments. An example from Hayonim cave is illustrated. D. Henry (1976) illustrates several other specimens from Hayonim Terrace.

A number of small bone slivers, about 30 mm long and pointed at both ends, have been found in Natufian sites. These have been identified as fish gorges ("gorgets") (Garrod and Bate 1937: 37), although this attribution is open to question. Several of these appear to have been made using the groove-and-splinter technique. No waste fragments proving conclusively that this method was used have been found, but the implements themselves show indications of split grooves on either side. These tools seem to have been made by cutting two parallel grooves a few millimeters apart through the shaft wall of a rather small long bone, possible that of a small mammal. The central splinter was then removed and its ends sharpened with a flint. In some cases this sharpening is confined to the very tips; in others the entire implement was worked, so that the original method of obtaining the fragment is no longer apparent. Some specimens were clearly made from irregular bone fragments. One specimen from Hayonim Cave is of particular interest[10]: the splinter was placed so as to include a nutrient foramen of the bone within it. The foramen runs axially part of the length of the artifact, terminating near its sharp tip. The function of this tiny channel may have been to contain a fluid, but there is no visible residue.

Plate 12: Small Double-Pointed Artifact (HAY 129) Made by "Groove-and-Splinter" Technique (c. 12X)

Natufian assemblages also contain a number of small barbed points, ranging from 43 to 150 mm in length. Several barbs were fashioned along one edge of the shaft. The manufacture marks indicate that these artifacts too were made using a flint edge. More elaborate work with a burin facet was needed to fashion the barbs themselves. The barbs were made by cutting notches alternately from either face of the point.

Several of the most striking art objects found in Natufian assemblages take the form of bone "sickle hafts". Hafts with simple cross-hatch decoration and undecorated hafts also occur. Haft fragments are commonly found. They are usually made from a section of the shaft of a large long bone. Marks on the surface indicate that they were largely formed by flint tools, although some roughing out by abrasion is possible. A deep groove, suitable for the mounting of flint blades, was cut into one edge of the bone haft, using a burin

---

[10] This specimen was not available for inclusion in the original study upon which this volume is based. As a result it is not included in the tabulations.

Plate 13: V-Shaped Groove on Edge of Bone Haft (WAD 21), Intended to Hold Sickle Blades; Note Notching, Presumably to Improve Adhesion of Cement (c. 10X)

which must have had a triangular cutting edge. These grooves have a deep V-shaped cross-section, suitable for receiving the typical Natufian bifacially-retouched sickle blades. These blades were fixed in place using some form of cement. Several specimens exist in which the blades remain in place. On one haft a cross-hatch of burin cuts was made in the bottom of the mounting groove, almost surely to improve the adhesion of the cement.

Burins with triangular facets were also used to cut decorative cross-hatch patterns on a number of bone hafts. These hatch lines are very straight and parallel, suggesting that they were made by guiding the burin edge against a straight-edge. Surface markings also indicate that the elaborate bone carvings for which Natufian is noted, such as the hafts from Kebarah and El Wad with representations of deer on the handles, were made using flint scrapers and burins.

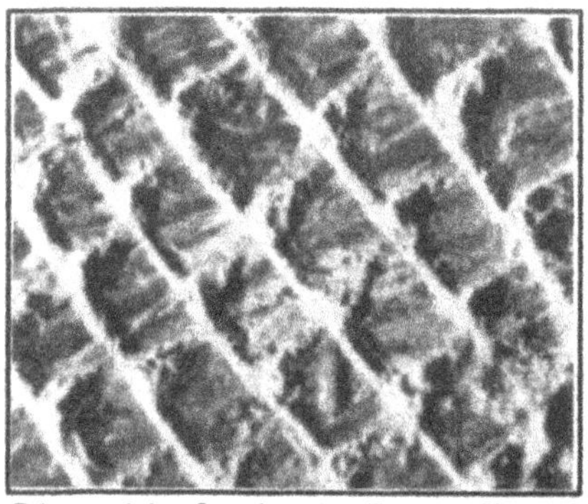

Plate 14: Incised Cross-Hatch Decoration (c. 10X) on a Bone Haft (HAY 159)

Many Natufian bone objects are perforated. The majority were likely perforated with a hand-held borer. The holes are irregular and almost always have the typical hourglass section which results from working bone from both sides. Bone pendants seem usually to have been perforated with a hand-held flint. Among bone implements, perforations were often made at the basal or handle end. Most of these holes were probably for the attachment of a suspension cord. Holes were commonly made through the distal epiphyses of pointed implements made on split metapodials. The hole was started in the natural depression on either side of the epiphysis. A few implements seem to have been pierced using some form of rotary drill. An implement from Shukbah has a markedly regular hole with parallel sides; on either side of the hole are a number of small regular circles which must have been false starts with the drill. Such drilling devices must have been scarce, however.

Plate 15: Perforated Implement (SHK 1), Showing False Starts by Drilling Device (c. 10X)

A very unusual implement from Kebarah makes use of the large natural foramen within the skull which occurs just beneath the horn core of the gazelle. The shape of the horn core itself is little modified, but the bone at its base was trimmed away, leaving the foramen as a perforation at the base of the implement. The base, with foramen, of what was likely to have been a similar implement was also found at Hayonim Cave. It is of considerable interest that such a specialized technique should have been used at two fairly wide-spaced sites.

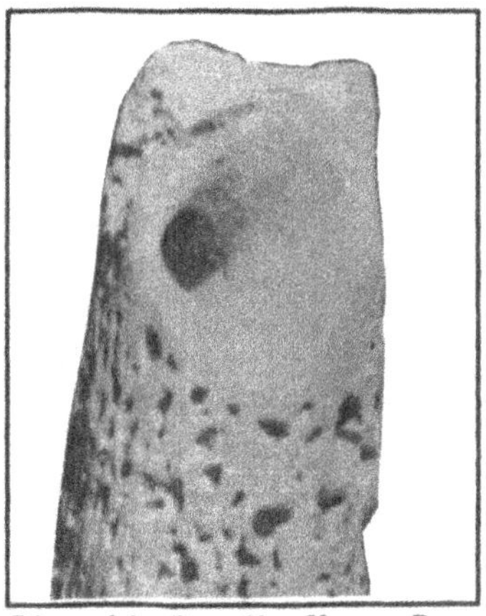

Plate 16: Gazelle Horn Core Implement (KEB 43) with Foramen Forming Perforation at Base (c. 2X)

Several hook-shaped objects were found at Kebarah Cave. These were fashioned by drilling a hole in a flattened piece of bone (in one case actually a tusk; enamel is visible on one side). Two parallel cuts were then made with the grain of the bone, from the hole to the edge of the fragment. The resultant U was then rounded and finished with a flint edge, and one case by abrasion, to make a point on one arm. A groove was cut around the longer arm, evidently to hold a cord.

Plate 17: Bone Hook (KEB 24) (2X)

This study has been confined largely the analysis of the techniques used to make Natufian bone tools. Natufian bone assemblages also contain numerous bone beads and characteristic bone pendants. These objects have not formed a central part of this study. Only a small sample of Natufian beads and pendants has been examined. Such decorative objects (which are quite common) appear to have been usually manufactured by abrasion and are frequently intentionally highly polished (see Stordeur, 1980). Bone implements, however, were seldom made by abrasive techniques[11].

The rare use of abrasive technique was not confined to any particular tool type. Most such implements do not differ substantially in form from other examples which have been shaved to shape. Included are a pointed implement on a complete distal metapodial with an axially ground tip and a pointed implement on a split distal metapodial with an axially ground tip and basal (epiphysis) end flattened and smoothed by cross grinding. A small bipoint was shaped by both axial and cross grinding, with the result that one of its tips is slightly spatulate. One very narrow point appears axially ground, and two very large implements with flattened

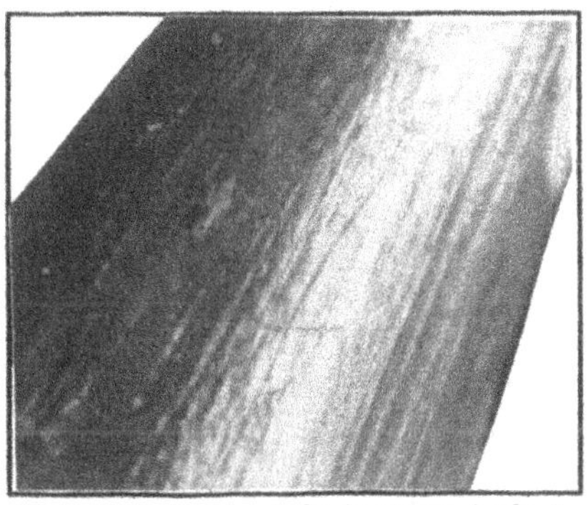

Plate 18: Typical Striations Left by Axial Shaving (c. 10X) on a Pointed Implement (HAY 61)

[11]In the comparisons to follow between the manufacture methods of the Natufian and Zagros Protoneolithic cultures, the reader should bear in mind that it is the bone tool assemblages that are being compared. Although the Natufian people were well aware of abrasive techniques they chose not to apply them to bone implement manufacture.

surfaces, both of antler, were also smoothed by abrasion. None of the Natufian implements included in this study shows clear intentional polishing. Some of the fragments show smoothed surfaces, but the method by which this was done is not known. One large implement from Kebarah, shaped rather like a large knitting needle, appears intentionally smoothed on the central shaft. Very fine, evenly spaced longitudinal scratches surround the central portion. These could have been produced by rubbing with a fine abrasive.

Zagros Protoneolithic
Manufacturing techniques

Formally the Zagros Protoneolithic bone tool assemblage bears much resemblance to that of the Natufian. Pointed implements and point fragments are the predominant forms. In addition the Protoneolithic assemblage includes a number of tools with broad spatulate tips. These are commonly made on complete long bones with the epiphysis as a handle. They may have served in wood or hide working. Small pin-like forms occur. A elongated implement with a flat cross-section may have served in basketry or weaving. A few bone hafts for flint blades have been found.

Because nearly all the Protoneolithic bone implements are made on shaft fragments from which the epiphyses have been removed, it has not been possible to assign a source to any but a few of them. As with the Natufian tools, small artiodactyls would provide long bones of appropriate size and shape for many of them. An unfused distal radius and two implements made from highly modified split distal metapodials are clearly from small artiodactyls, but could not be more closely identified, as is the case for a spatulate implement and a basal end made from split distal metapodials of large artiodactyls. There are four antler tine tips in the sample. One spatulate implement and one pointed implement are made from the distal tibiae of *Dama*.

Plate 19: Axially Ground Zagros Protoneolithic Pointed Implement (SHN 22) (½X)

The surfaces of most implements are highly worked. Consequently it is difficult to determine the means used to obtain a bone blank. Only three specimens, two points and a pointed implement, still show traces of incised grooves on the edges. These suggest a grooving-and-splitting technique similar to that used by the Natufians. It is probable that numerous others were similarly sectioned, but the traces have been obliterated. The few tools made on complete long bones appear to have been twist-fractured. Most blanks seem to have been obtained by simple shattering of the bone and the selection of appropriate fragments. There are many darkened fragments, so heating may have been used to help split the bone shafts longitudinally.

Almost all the specimens have been ground to shape by rubbing against an abrasive stone of moderate coarseness.

Most specimens were first rough shaped by grinding with a back-and-forth movement, at about a forty-five degree angle to the axis of the shaft. Points were then sharpened by axial grinding. There are many implements with spatulate rather than pointed tips; these tips were shaped primarily by cross-grinding, leaving flats and distinct facets. A few specimens were axially shaved to a point with a flint edge, much as in the Natufian assemblage.

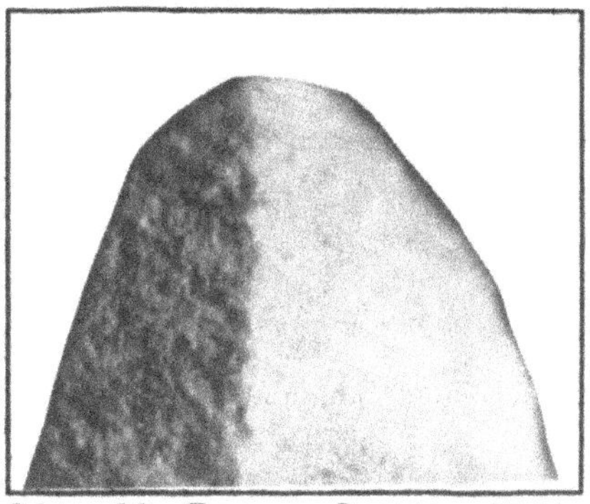

Plate 20: Zagros Protoneolithic Spatulate-Tipped Tool (ZCS 82) Showing Cross-Grinding (c. 10X)

A very common artifact type found at Zawi Chemi is a bone fragment which has been axially ground to a point. The very tip was then scored around with a burin or flake and the tip snapped off. Few of these objects show any wear at the broken-off tips; they are not likely to have been tools themselves. Two small pin-shaped bone objects were found at Zawi Chemi Shanidar. These tiny objects (16 - 17 mm long) are sharply pointed at one end and wider at the other. One specimen shows clear traces of score marks around its base. These pin-like objects show signs of use-wear. It appears very likely, therefore, that the artifacts with broken-off tips are waste fragments left in fashioning the small pins. This technique contrasts sharply with the groove-and-splinter technique used by the Natufians to make the similar sized small bipoints.

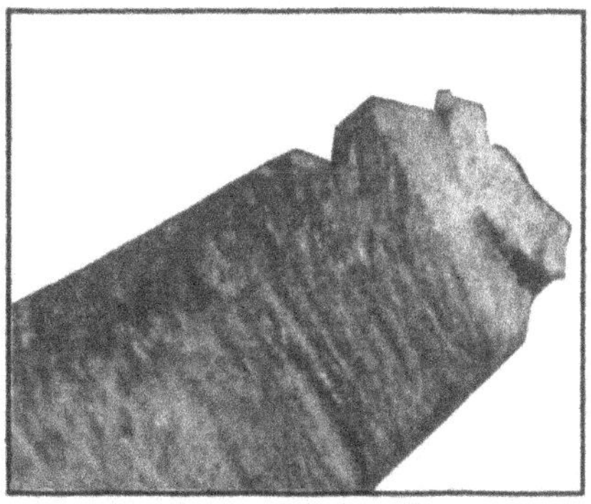

Plate 21: Cross-Ground, Scored, and Snapped Tip (ZCS 110); A Waste Product From the Manufacture of Tiny Bone Points (c. 10X)

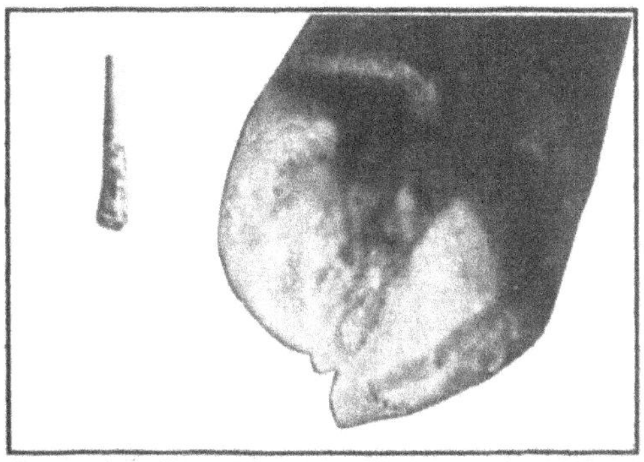

Plate 22: Tiny Bone Point (ZCS 4), Probably Snapped From Blanks Such As That Above; (Base, c. 10X)

Two hafts were found at Shanidar Cave and Zawi Chemi Shanidar. Although these otherwise bear little formal resemblance to the Natufian equivalents, the method of making the groove at the edge seems virtually identical. A flint implement with a burin-like facet was almost surely used to cut the deep V-shaped grooves in which flint blades were hafted. Burin-like facets were also used to cut the incised decoration seen on numerous Protoneolithic bone artifacts. Fully rounded carvings, however, have not been found.

Perforations have been made in several Protoneolithic implements. These holes generally have parallel sides and are quite true, suggesting the use of some sort of rotary drilling device. All the examples, however, are either somewhat worn or broken, so it is not possible to reach a positive conclusion.

Two very small broken holes, bored at the end of very small implements show a distinct hourglass shape, indicating boring from both sides. These must surely have been made with a hand-held borer.

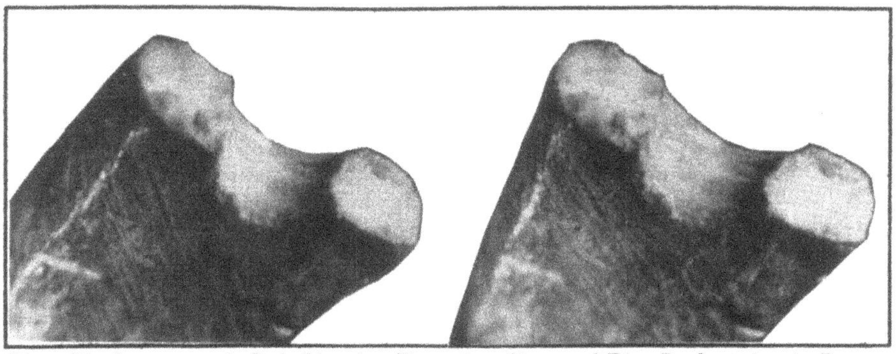

Plate 23: Stereoscopic Pair Showing Hourglass Shape of Tiny Perforation in Zagros Protoneolithic Implement (ZCS 148); Note Also the Cross-Grinding (c. 10X)

There is little indication of intentional polish on the implements, although some of the decorative bone objects may have been smoothed. The method used is not apparent.

CHAPTER IV

The Natufian and Zagros Protoneolithic Bone Tool Assemblages

Introduction

This chapter will briefly introduce the major categories of bone tools that will subsequently be described in detail. No attempt has been made to subdivide the tools according to a strict formal typology. Rather, they are broken into groups which appear sufficiently similar that they might potentially have served similar functions.

In this study a simple terminology will be used to describe the various attributes of bone tools. Although the terms "distal", "proximal", "medial", and "lateral" are in common use to describe bone implements, I have intentionally avoided them here. My intent is to avoid confusion between bone tool landmarks and the anatomical terminology that is applied to the bone from which the tool has been made. The working end of an implement is designated as the "tip"; the handle or hafted end is referred to as the "base". If the shaft of an implement is wider than it is thick the broader surfaces are called "faces" and the narrower surfaces are called "sides." Terms such as "distal," "proximal," etc. always refer to the anatomical landmarks of the bone from which the tool was made.

Points and pointed implements

Points and pointed implements are the most numerous forms to be found in both the Natufian and Zagros Protoneolithic assemblages. These are tools made on long bones or elongated bone fragments. One end has been tapered to a point. The tip itself may not be sharp; most tips are irregularly hemispherical. Incomplete pointed bone fragments will be referred to as points. Many of these are quite small, and are likely to be broken-off tips of larger implements. Others may actually be complete but show insufficient evidence (such as wear) at the base to indicate that the break was not accidental. Some points may have been hafted, although there is little evidence for this. Tools designated as pointed implements appear to be complete in themselves or nearly so. The basal end is usually present or the wear on the existing portion indicates that the implement was hand-held. The forms of most pointed implements makes it unlikely that they were hafted in any way.

The Natufian pointed implements may be subdivided into several groups. The size and shape of an implement depends in large measure on the anatomical part from which it was made. Implements made on complete

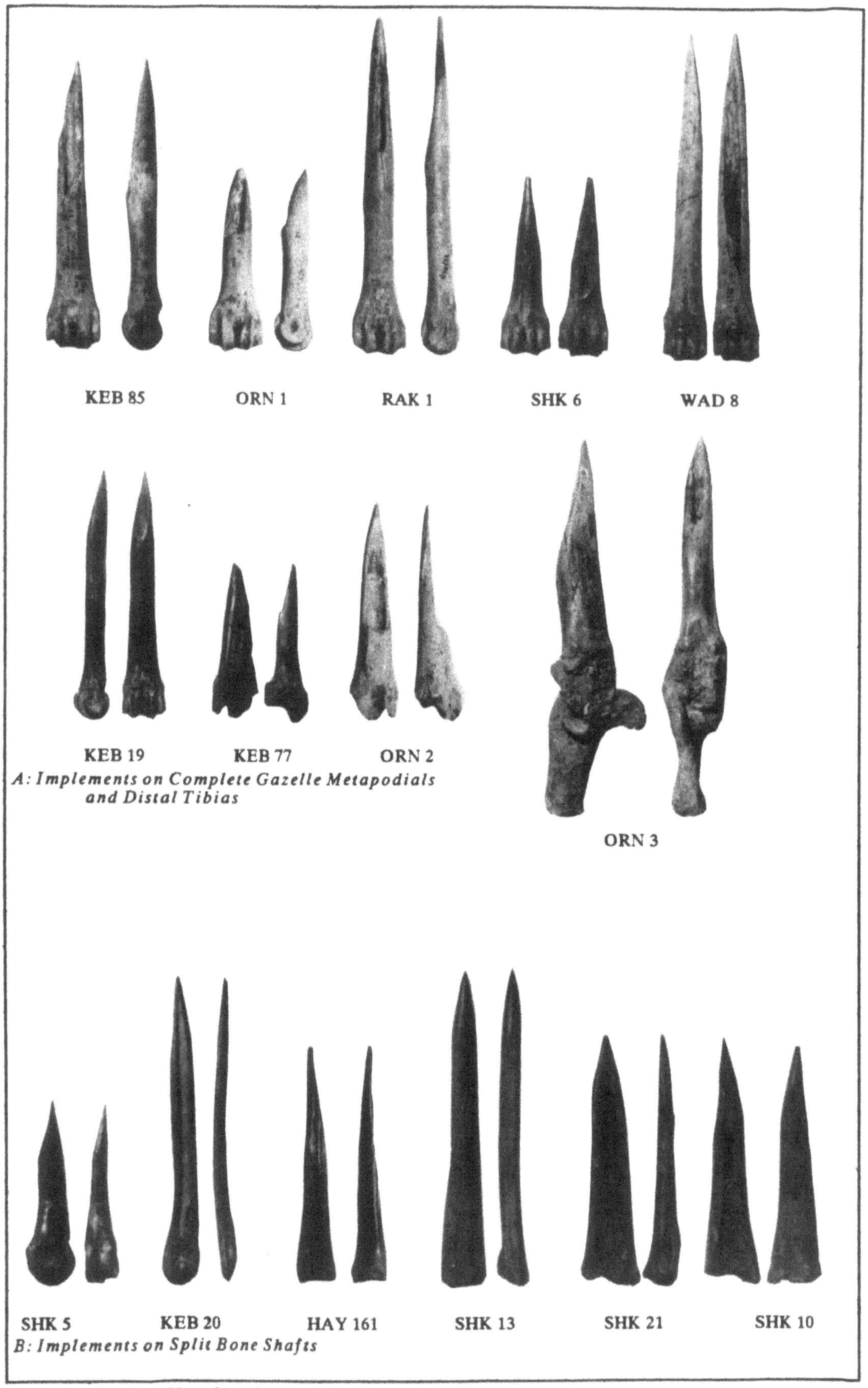

A: Implements on Complete Gazelle Metapodials and Distal Tibias

B: Implements on Split Bone Shafts

Plate 24: Typical Natufian Pointed Implements (¼ X)

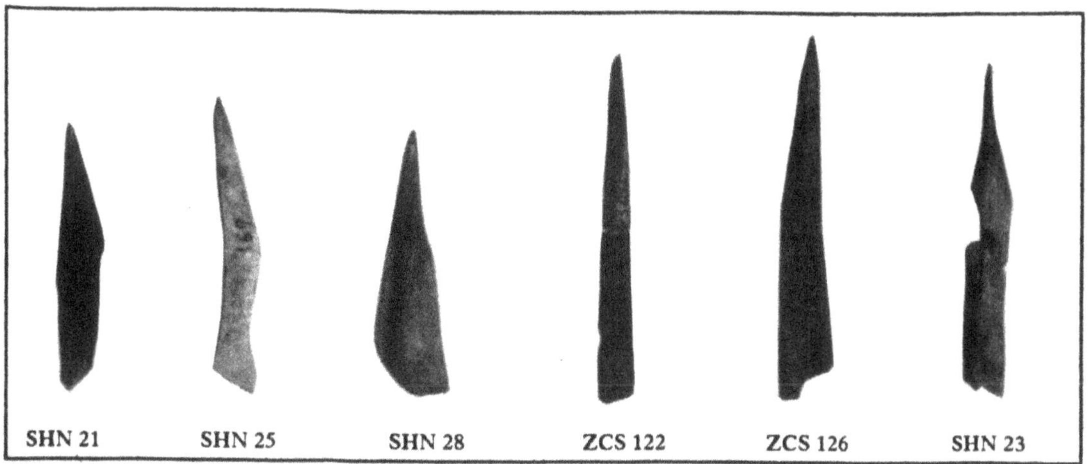

Plate 25: Typical Zagros Protoneolithic Pointed Implements (¼ X)

proximal and distal metapodials are the most robust. Four of the tools made on complete long bones of other types also fall into the most robust group. Other implements are made on metapodials and long bones which have been split lengthwise. These tools are smaller and lighter. Those made on shaft fragments vary widely in size. The pointed implements made on shaft fragments do not form a single coherent grouping. With the exception of three specimens, which are much larger than the others, these tools fall into a continuum of forms and sizes which is difficult to subdivide. The majority of these implements have faces and sides which are parallel or nearly parallel to one another. A number of similarly shaped incomplete points clearly belong to this group.

The points and pointed implements from the Zagros Protoneolithic sites present a simpler picture. With the exception of a tool made on an unfused radius, all specimens with pointed tips are made on rather robust shaft fragments. Shaft shapes range from a rounded to an irregular cross-section. The tools characteristically taper from the irregular shape of the bone fragment to a round cross-section at the tip. Consequently points broken off near the tip often have this shape. There is no indication that they represent a separate form. As there is no clear criterion for subdivision, the Zagros Protoneolithic points and pointed implements will be considered together as a group.

Natufian double-pointed implements

Natufian sites have produced a number of bone implements similar in that they are all narrow objects pointed on both ends. They are of greatly varying sizes. These implements may be divided into a clear group of small double-pointed objects (usually termed "gorgets") and a less well-defined group which varies in size from only slightly larger than the small double-points to a very large implement 177 mm long[12]. The complete specimens of this last

[12]When this study was initially conducted, a third group of relatively large, thick implements, termed "bipoints", was also included. These now appear to be Aurignacian in origin.

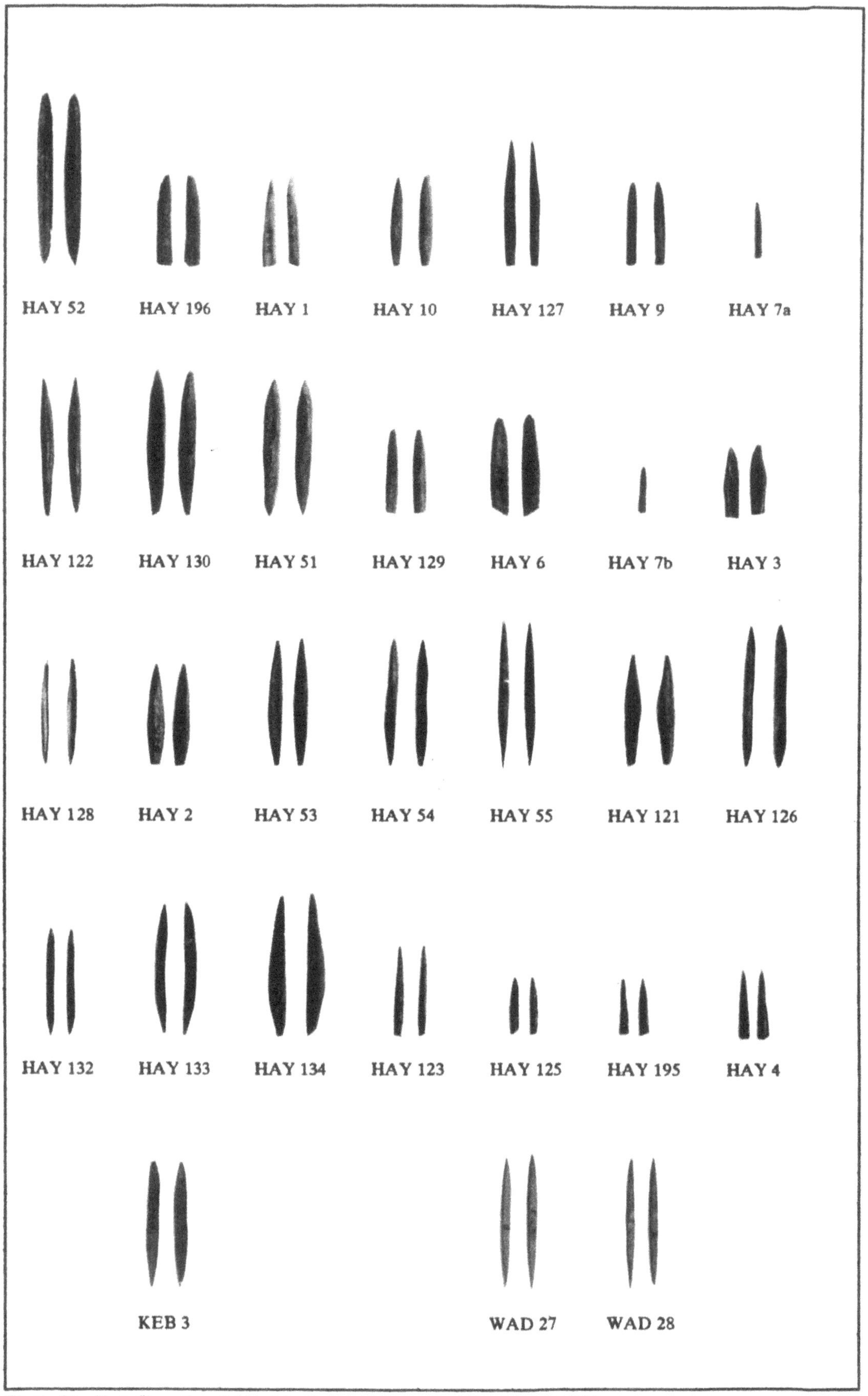

Plate 26: Natufian Small Double-Pointed Bone Artifacts (c. 3/4 X)

group are characteristically quite narrow in comparison to their length. In addition there are a number of broken specimens.

The small double-points were worked into final shape by scraping with an edge, presumably of flint. The action was much like that of sharpening a pencil with a knife, resulting in sharpened ends. This sharpening action may have been confined to the very tips of the object, leading to specimens with nearly parallel sides, or may have been carried back to the midpoint, leading to convexly curved sides. The original blank may have been an irregular fragment or slip of bone, often from the shaft of a small long bone, or it may have been intentionally cut from the shaft of a small long bone. There are some specimens which are too heavily modified in the final sharpening to be certain of the origin of the blank. These that were intentionally cut were extracted from the bone shaft by the "groove-and-splinter" technique (see Chapter III).

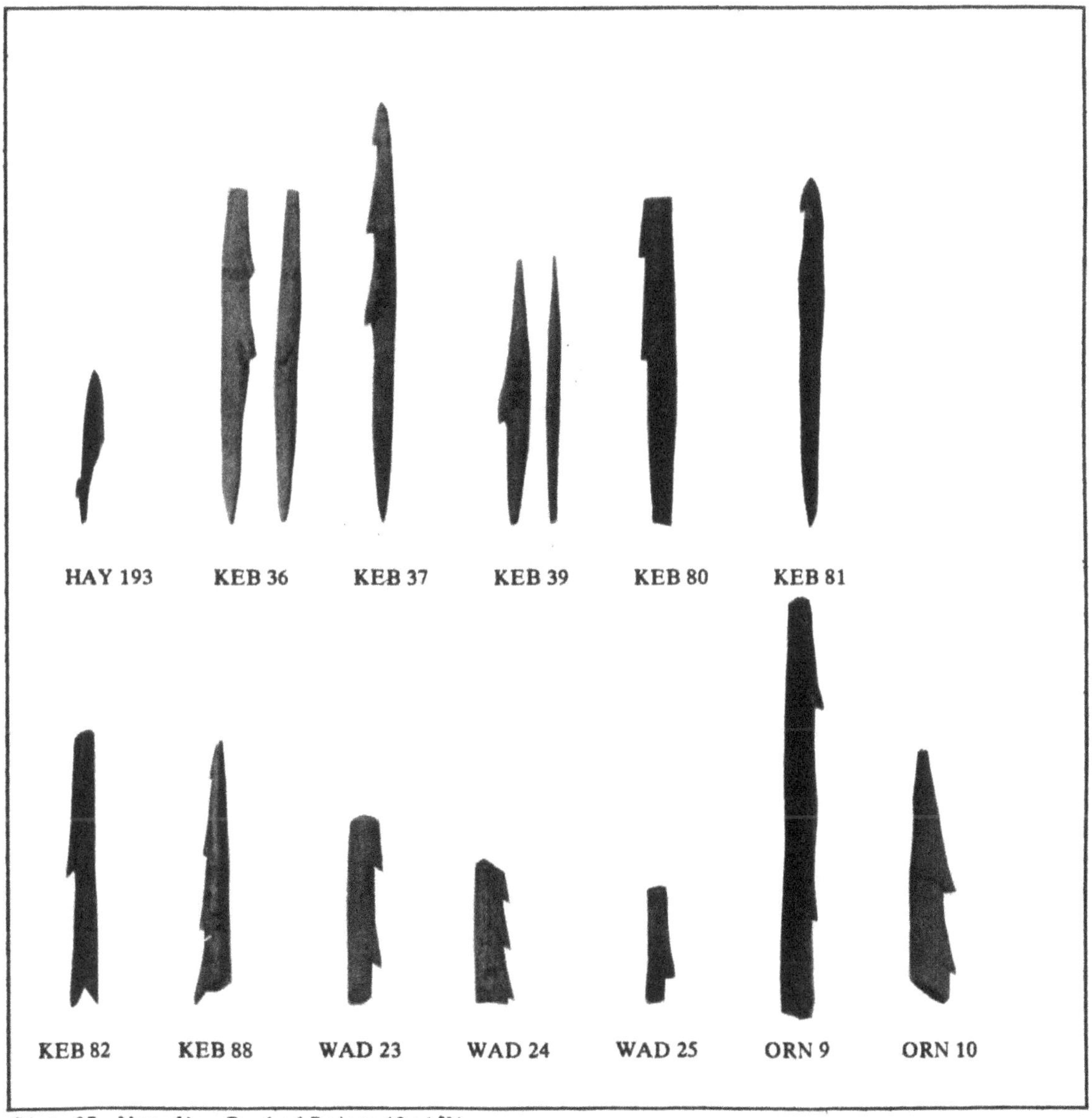

Plate 27: Natufian Barbed Points (3/4X)

Natufian barbed points

Points bearing a single row of barbs along one side have been found at several Natufian sites. These are usually referred to as "harpoons," and interpreted as projectile points.

Spatulate-tipped implements

This category actually consists of a number of dissimilar forms. Generally it includes tools on which the portion contacting the worked material is broad, rather than pointed. This broad working portion is usually shaped somewhat like a spatula (or perhaps better, like the working end of a chisel). Other specimens have very large working areas which extend well beyond the tip ends, sometimes covering much of the area of the tool.

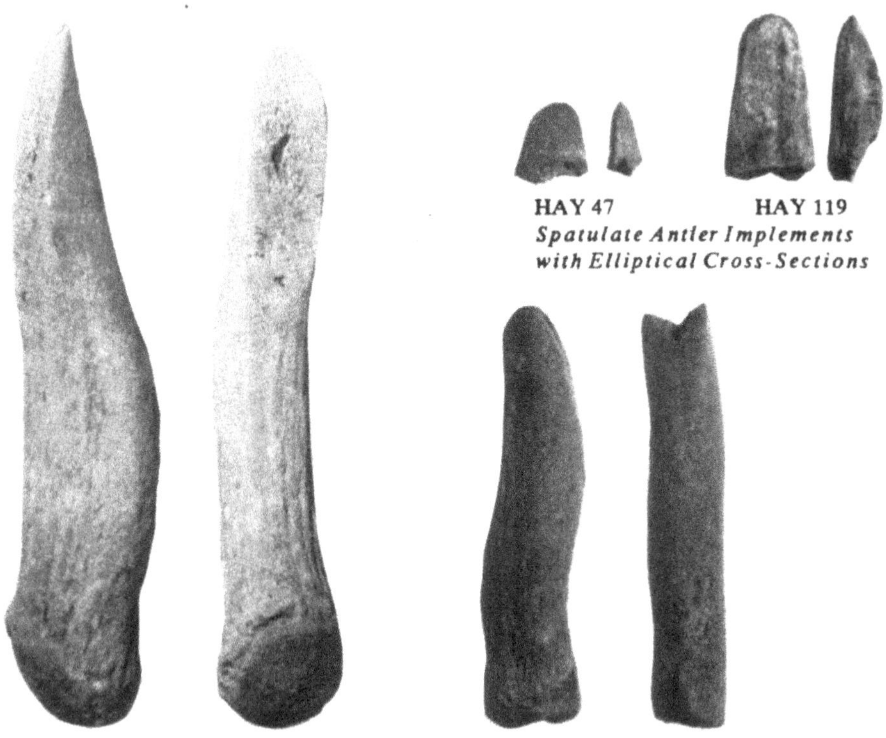

HAY 47 HAY 119
Spatulate Antler Implements with Elliptical Cross-Sections

WAD 14 HAY 191
Very Large Antler Tools with Single Flattened Surface

Plate 28: Large Natufian Spatulate Implements of Antler (½ X)

The group may be subdivided into (1) large implements, mostly of antler, with large, robust spatulate tips with elliptical profiles; (2) large to very large antler implements with a smoothed and polished flat surface on one side cut diagonally across the shaft so as to form a roughly spatulate tip, although the shaft is otherwise unworked; (3) bone implements, on shaft fragments or complete long bone shafts, with intentionally made spatulate tips of various sizes; (4) a group of long, flattened implements, often made on ribs, which are generally bodkin-like. Some of these have perforations; some have spatulate tips while others have round tips.

ZCS 92	KEB 27	SHN 18	SHK 4	WAD 9
Spatulate-Tipped Tool (Zagros Protoneolithic)	*Spatulate Tool (Natufian)*	*Perforated Flattened Implements (Protoneolithic)*		*(Natufian)*

Plate 29: Typical Spatulate-Tipped and Flattened Implements (½ X)

Zagros Protoneolithic snapped-tipped objects

Twenty-seven small bone objects were found at the Zawi Chemi Shanidar which are superficially similar in shape to bone points. Each object, however, had been scored around the tip end and the very tip was then snapped off. One specimen is double ended, with both tips snapped off. These are evidently waste products (see Chapter III).

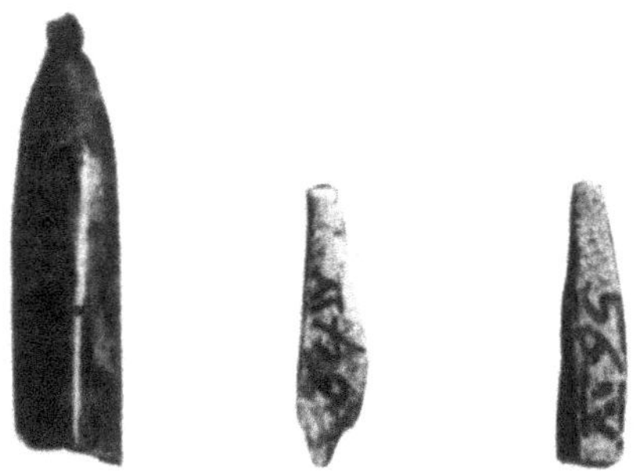

ZCS 112 ZCS 138 ZCS 141

Plate 30: Typical Zagros Protoneolithic Snapped-Tipped Objects

Pin-shaped implements

A few pin-like objects occur in both the Natufian and Zagros Protoneolithic assemblages. The Zagros Protoneolithic objects appear to be the desired product for which the snapped-tipped bone fragments served as blanks. HAY

ZCS 4 ZCS 14 HAY 120 HAY 124

Plate 31: Typical Pin-Shaped Implements

124 is remarkable for the care taken in its manufacture. It is very regular, with three distinct raised ridges at its center.

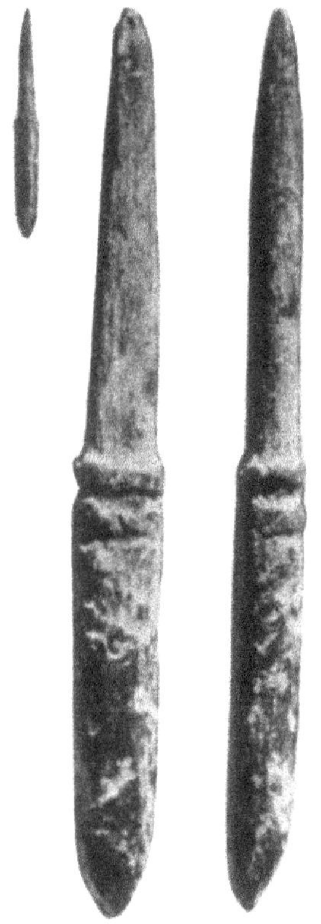

Plate 32: Pin-ShapedObject HAY 124 (1X & 4X)

Flattened implements

Both Natufian and Zagros Protoneolithic assemblages include a few miscellaneous objects similar only in that these elongated implements have a markedly flattened cross-section.

Bone hafts

Bone hafts are found at both Natufian and Zagros Protoneolithic sites. These are usually made from a section of the shaft of a large long bone. They are elongated implements, usually with a flat cross-section and a V-shaped groove on one edge into which the flint blades had been inserted (see Chapter III). It is certain that flint blades were attached in this manner as several specimens exist in which the blades remain in place. These implements have usually been interpreted as "sickle-hafts." Sickle blades, bearing the characteristic"sickle-gloss," which results from cutting the siliceous stalks of grain or reeds, are common on Natufian sites. The blades mounted in the hafts may have been sickle blades.

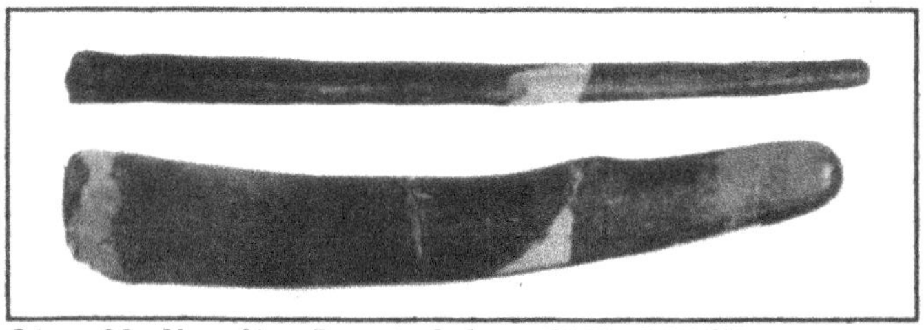

Plate 33: Natufian Bone haft from El Wad (½ X)

Toothed object

A flattened bone artifact bearing a row of short teeth along one edge was found at Kebarah Cave. Turville-Petre (1932) considered this object a comb. (See Page 106).

Natufian perforated scapula

A perforated deer scapula was recovered in 1928 by C. Lambert from a trial trench in the cave of El Wad. This implement was briefly described by Garrod (1930) and attributed to the Natufian. In a later publication this implement is illustrated as belonging to El Wad Layer B2 (Lower Natufian)(Garrod 1932). Garrod remarks that the tool recalls the *baton-de-commandement* of the Western European Upper Paleolithic. (See Page 107).

Antler tines

A number of antler tines have been recovered from both Natufian and Zagros Protoneolithic sites. The majority of these are of red deer antler. A specimen from El Wad is a roe deer antler tine. These tines have simply been broken off or the shaft has been notched and the tines then broken off. They are not worked in any other way.

Miscellaneous objects

This category includes fragments, the possible handle portion of tools, and other uncommon objects that are difficult to classify or interpret.

CHAPTER V

Analysis of Use-Wear: Ethnographic and Experimental Evidence

Points and pointed implements

Potential uses for points and pointed implements

Points and pointed implements, owing to the simplicity of their shape, may have served a variety of purposes. These uses, however, all share a basic similarity. The only likely purpose in making a sharpened or narrowed tip on such an implement is to ease the penetration of the tip through some particular material. Among the possibilities:

1. Those points which would have been suitably shaped for hafting may have served as projectile points. Those points with sharpened tips could have been used to penetrate an animal's flesh, but blunter tips could have been used to stun smaller game.

2. Sharply-tipped implements may have been used as perforators for relatively dense materials such as hides. As such they would have been part of a leather-working kit. Some limited use of such tips for wood-working is also possible.

3. Implements with tips too broad to have penetrated hides may also have been used in manipulating materials or fabrics which would admit the tip readily. Implements of this sort may have been used as manipulators in the related crafts of weaving, netting, mat-making, and basketry.

4. Some pointed implements may have been used as retouchers in the manufacture of lithic artifacts.

Bone awls of various forms are well-known from ethnographic contexts, particularly among Native Americans. These tools were used both as perforators in the working of skins and leather and as manipulators in the making of basketry. Indeed, the same tools were often used interchangeably for both purposes. Bone perforators appear to have been used by virtually all tribes who regularly worked hides, although descriptions of the use of such tools are few. Clothing, tents, and other skin objects were usually joined by making appropriate perforations with a bone awl and lacing with sinew (see, for example, Grinell 1901 and Wissler 1910 on the Blackfoot). Very thick rawhide, such as that used in making "parfleches" (saddlebags), could be perforated by burning-in holes with the heated tip of such an implement

(Morrow 1975). Such a use may indeed be the best explanation for the occasional specimen with a heated tip found in both Natufian and Zagros Protoneolithic assemblages.

Tribes that made baskets (such as the California tribes) used similar implements to separate the basketry coils so that the binding cordage could be inserted. The slightly rounded points of such tools were superior to sharply-pointed metal tools, as they would separate the coils without splitting apart the fibers (Mason 1904: 246). Such basketry tools seem to have been ubiquitous, but basket-makers generally also worked hides:[13] "In every living tribe of basketmakers these (bone) awls are among the commonest of women's tools, most serviceable in sewing garments as well" (Mason 1904: 246).

For projectile points and perforators the primary working portion of the tool is the tip itself. Tools used as manipulators have their primary contact with the worked material in the area surrounding the shaft and extending back from the tip, depending on the nature of the use. A bone projectile point is propelled at its target, which it may be expected to penetrate with little or no rotation and with considerable force. If the point scores a direct hit it may penetrate soft flesh; more often, however, it may be expected to strike unyielding materials such as bones, trees, or stones, or to be thrust into the soil, which would likely contain abrasives such as gravel and sand. With this in mind, E. E. Tyzzer (1936) conducted a series of experiments in which he regularly shot a number of bone points into an earthen bank, using a bow. The wear observed on the points was much as would be expected: the penetration of the point into the abrasive soil resulted in the smoothing and rounding of the tip and the polishing of the tip area. With repeated shots, however, the tip will contact stones and pebbles and eventually become battered and broken. Tip breakage of this type presents an irregular, shattered appearance and must be distinguished from the smooth, regular, and usually angled break which commonly results from sideways pressure.

For this study a number of experiments were conducted using bone points as perforators for leather and hides. Bone points may be used for this purpose easily and efficiently provided that the tip diameter is small. Thin, delicate materials may be pierced by steady hand pressure; the penetration of thicker materials may be greatly improved by twisting the tool (rotating it about its axis).

Presumably, if a point is intended as a perforator the tip will be made as sharp as possible. Experiments in bone tool manufacture indicate that sharp tips are easily obtained both by flint shaving and abrasive techniques. A trade-off is involved however. Sharp tips are more easily broken than

[13]Perhaps because such tools were so commonplace, few specimens have found their way into museum collections. Of these very few are adequately documented as to use; many may well have has multiple uses. Consequently the usefulness of such specimens as comparatives is limited.

broader ones; consequently while sharp tips may be best for working fine materials they will quickly break when used on thicker, more resistant hides. The optimum tip size depends both on the nature of the work and the skill and care of the worker. It is difficult to estimate the maximum tip size for an effective perforator as this is dependant on the amount of force the worker is willing to exert. Any worker, however, would be likely to avoid unnecessary effort. Blunted tips can easily be resharpened.

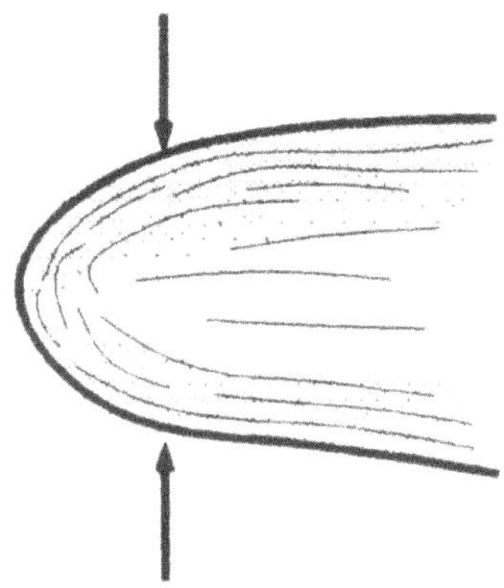

Figure 3: Measurement of Tip Diameter

Based on experiment a reasonable maximum tip diameter for a perforator may be estimated at 1.0 to 1.5 mm with the tip diameter estimated as illustrated in Figure 3. A brief experiment illustrates this point more rigorously. Bone points with tip diameters of 0.4 mm, 0.7 mm, 1.0 mm, and 2.0 mm were prepared by intentionally rounding the tips of pointed implements (made by abrasive methods) to the required dimensions. This was accomplished by careful rounding of the tip to a hemispherical shape with a fine abrasive stone. Samples of rabbit pelt and leather (tanned cowhide) split to various thicknesses (2.7 mm and 0.7 mm) were firmly attached to the platform of a spring scale. The leather and skin samples were backed with a

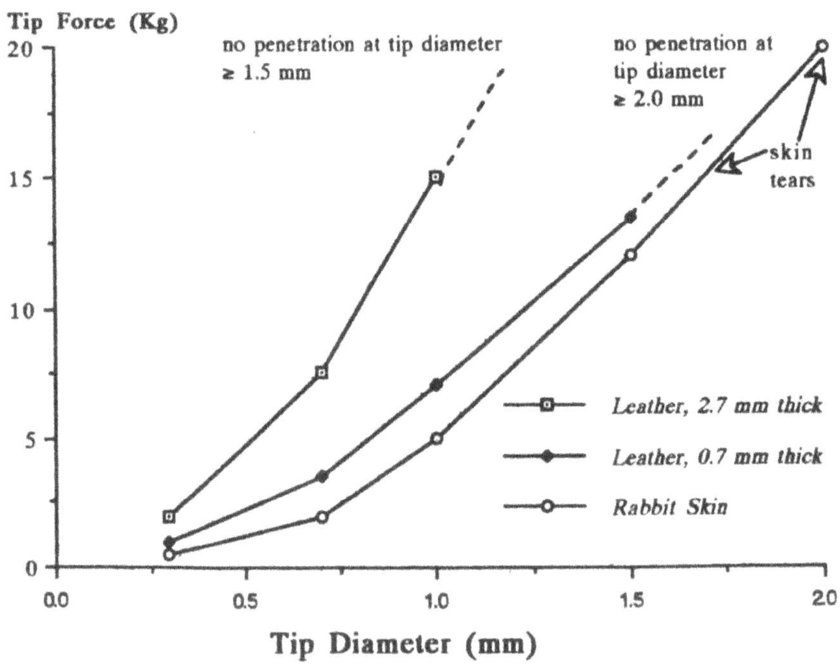

Figure 4: Pressure Required to Force a Bone Point Through Leather or Skin

thick cloth pad. A series of attempts was then made to pierce each sample using each of the different-sized bone points. Six to eight attempts were made with each leather-tool combination. In each case the pressure exerted by the tool-tip on the leather surface was measured just before penetration was achieved. The reader is referred to Figure 4, which illustrates the results of this experiment. It should be noted that with 2.7 mm thick leather and a broad tip of 1.5 mm and with 0.7 mm thick leather and a tip diameter of 2.0 mm penetration could not be achieved at all. When such a tool-tip was used on the very thin rabbit pelt tearing resulted.

Since considerable effort is exerted on the tip of a perforator, this area rapidly becomes rounded and polished by abrasion. The irregularities of the fresh tip are reduced and the fine striations left from the manufacture process gradually are obliterated, leaving the surface of the bone smooth and polished. This rounding and polish may extend some distance up the shaft from the tip, depending on the amount of use and how the tool penetrated the workpiece. The rounding and polish generally diminish gradually moving up the shaft from the tip, so that no clear line of demarkation is visible. Experiment has shown that the effect of leather and hides by themselves is to smooth and polish the bone surface rather than to produce easily visible scratches. If the bone surface is worn very smooth and highly polished, however, fine striations may be visible as patches of fine, straight, parallel lines crossing the polish. Optical effects make extremely fine scratches more visible than they otherwise would be.

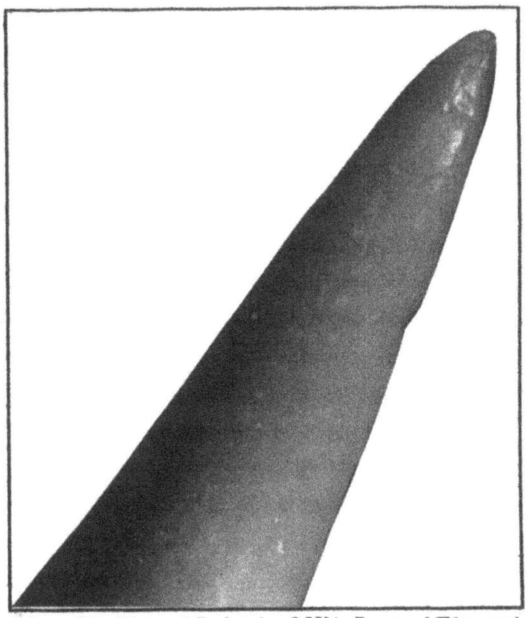

Plate 34: Shaved Point (c. 25X), Pressed Through Clean Leather 5000 Times (Moderate Rounding and Polish, No Scratches)

If abrasive materials such as silt and sand are present on the worked material readily visible scratches may be left. Longitudinal scratches are difficult to discriminate from the marks of manufacture and the natural grain of the bone. They can usually be detected only of the bone is worn smooth. If the tool is twisted as it is used, however, readily visible scratches will be left. These surround the shaft concentric to the tip. They may run around the tip or spiral back from the tip at an angle. Presence of scratches of this type may be taken as proof that the implement was used as a perforator. However, as their formation is dependant on the presence of a foreign abrasive on the worked material, the lack of such scratches cannot be considered as evidence that a tool was not a perforator. Experiment has shown that unless an additional abrasive is present no patent scratches will be produced (except under the conditions of extreme

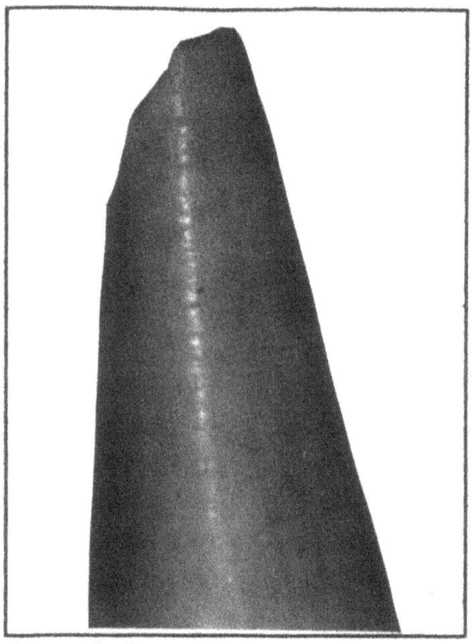

Plate 35: Shaved Point (c. 25X), Pressed 50 Times with a Twisting Motion Through Leather Sheet Rubbed with Soil (Broken Tip, Light Rounding and Polish, Many Light Rotation Scratches)

smoothing and polish mentioned above). Conversely, if an abrasive is intentionally introduced into the perforation being made in the leather, clear scratches will always be produced within a few strokes of the implement. Therefore, if a group of implements has been used as perforators of leather or hides the proportion of these specimens which will show diagnostic scratches will be directly related to the amount of abrasive sand and silt present. Unfortunately, there can be no way to directly determine this quantity. It is not likely, though that these hides were either very dirty or very clean. As the area of the hide contacted by the implement is quite small it seems likely a tool tip would contact an abrasive particle only rarely. A large tip or a tip pushed farther into the material would have a proportionately greater chance of encountering an abrasive and becoming scratched.

If a bone point is used to work a material markedly harder than itself the wear appearing upon the tip is massive. Six experimental points were used for this purpose. The use of the sharp tip of the experimental bone point to retouch a sharp flint edge quickly resulted in tip breakage and transverse scores across the broken end. The flint edge was left with a series of fine semicircular notches; the edge could then be easily regularized and straightened by using the side of the tool tip. This resulted in a series of transverse scores across the shaft near the tip.

Tools used as manipulators in weaving, basketry, etc., would make their principal contact with the worked material in the area immediately behind the tip. It is more difficult to characterize the expected wear on these tools because the manner of their use is more variable. If such a tool must be forced into the worked material it may be twisted in the manner of a perforator, perhaps leading to concentric scratches of the type described above. For instance, a tool used to force a binding cord into bundles of basketry reeds might be used in this way. Because of the more open nature of the worked material, however, such concentric scratches are likely to be formed farther back on the tool shaft rather than at the very tip. If, on the other hand, the tool was used to press down the weft in weaving, matting, or basketry the tip may be left bearing scratches which run transversely across the tip but not around it.

Material like reeds are more abrasive than leather (as they contain silica) and can cause greater wear. The area of the tool contacting the worked material is many times greater than the contact area of a perforator. Because tool pressure and the consequent friction is distributed rather than concentrated at the tip, a manipulator would have to be in use a very long time to accrue a similar degree of rounding and polish.

Spatulate-tipped implements

Potential uses for spatulate-tipped implements

This diverse group of tools may have served a number of purposes. Basically, however, these functions fall into two categories: the use of the broad surface of the implement as a smoothing tool, such as for hide-smoothing, and the use of the edge of the tool as a wedge, chisel, or pry.

Experimental tools were manufactured and used during the course of this study in order to test the forms of wear to be expected from rubbing a tool across the surface of a hide and from the working, splitting, and shaving of wood.

Steinbring (1966) describes the manufacture and use of bone defleshing tools by the Ojibwa Indians. This tool was made from the immature metacarpal of a moose. These tools have a broad spatulate tip with an edge at right angles to the bone shaft. The blade of the tool was notched across its width. After the hide was slightly rotted to loosen the hair the excess tissue on the inside of the hide was removed with this implement. This bone implement seemed superior to a steel tipped tool also used (to remove the outer hair) as it did not penetrate too deeply into the hide surface. It is of interest that these implements remained in use for very long periods of time. One implement observed by Steinbring had been in use for eleven years; another specimen known to him had been in use for twenty-six years. The number of tools in use at any one time was very low -- about one tool for every worker in hides.

An experimental tool was made in order to simulate the wear expected on a tool tip from prolonged friction with leather or hide that would occur if the implement were used as a hide compressor, dresser, or rubber (Plates 35 and 36). This implement was ground to shape from a *Bos* long bone fragment. It had a spatulate tip about 10 mm wide, with a sharp tip edge at right angles to the axis of the shaft. This tool was mounted on a wear-producing apparatus consisting of a reciprocating motor-driven arm, hinged in such a way as to simulate as closely as possible the motion of the human arm. This arm moved the tool approximately axially back-and-forth with a 10 cm stroke about 2.5 times a minute. The arm was initially weighted to produce 1 kg of pressure at the tool tip. The tip was arranged to lie flat against an unfinished leather belt, which was mounted on a slowly rotating drum to constantly provide the tool with a fresh surface to rub against. The belt was padded to allow the surface to yield slightly to the tool pressure. This mechanical arrangement was chosen over a manual experiment because it permitted a

Plate 36: Experimental Ground SpatulateTool (c. 2½X), After Prolonged Rubbing Against a Leather Surface (See Text)

better quantification of the amount of wear to be expected under controlled conditions (pressure, stroke length, etc.), and because it produced an amount of wear in a reasonable time which would be equivalent to a very long period of use. The conditions of the experiment were chosen to roughly approximate use by an average individual.

Obviously such an experiment cannot be expected to be realistic. The state of the hide under actual condition could be very variable, ranging from greasy to dry, with or without hair, or containing variable amounts of gritty materials. These limitations must be taken into account in evaluating the experimental results.

After 1500 strokes of the arm the tool was examined. It showed an overall polish on the flat tool tip, but the marks of manufacture were still plainly visible and there was no marked rounding of the edge. After 5000 strokes the tip of the tool was markedly polished and the edge was slightly rounded. The weight of the tool tip was then increased to 2 kg and an additional 1500 strokes made. The tip was now observed to be highly polished and the marks of manufacture nearly obliterated. Faintly visible fine scratches were apparent in the polish under the microscope (c. 50 diameters). These led back at right angles from the tip and were parallel with the axis of the implement and the direction of motion of the arm. These fine scratches were uniformly spaced, of the same overall depth, and parallel. After an additional 3500 strokes the rounding and polish had increased only slightly but the fine scratches had become quite clear. An additional 5000 strokes (totaling 10,000 strokes at 2 kg and 5000 strokes at 1 kg) left the appearance of the tip unchanged.

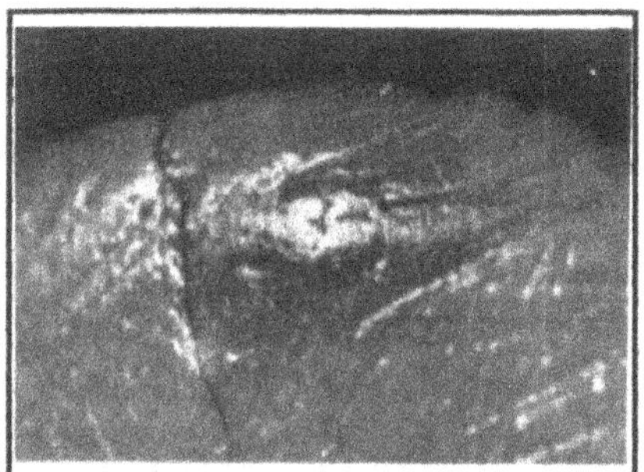

Plate 37: TheExperimentalToolAbove (c. 6X), High Polish, Rounding of the Edge, and Many Fine Parallel Scratches

Under actual conditions of use the depth and length of scratches and the degree of polish produced on the tip of such a tool may be expected to differ according to the conditions of lubrication and cleanliness of the hide. The general disposition of the rounding, polish, and scratches should be much like that on the experimental specimen.

Other experimental tools were used to test the wear to be expected if spatulate bone tools were used to work wood. The first of these was used as a wedge. This tool was ground to shape from a fragment of the shaft of an immature *Bos* radius. This tool was 62 mm long, 25 mm wide and 10 mm thick. The break at the base was at approximately a right angle to the shaft. The tip was ground to a sharpened spatulate edge about 9 mm across.

Initially this implement was driven into the end (into the grain) of a pine block twenty-five times, penetrating the wood about 1 cm. This entailed about 250 blows on the basal end. A *Bos* radius was used as a mallet for this purpose. The implement was then examined and photographed. The tip was very slightly rounded and polished but the marks left from manufacture remained clear. There were no chips or fractures. The butt end appeared slightly battered and the sharp ends of the bone had been reduced. There were no fractures on the base. The implement was then driven into the block an additional 75 times (approximately 1000 blows). The tip became slightly more polished but no fractures appeared. A number of flattened facets, having a slightly more polished appearance, appeared on the base from the blows of the mallet.

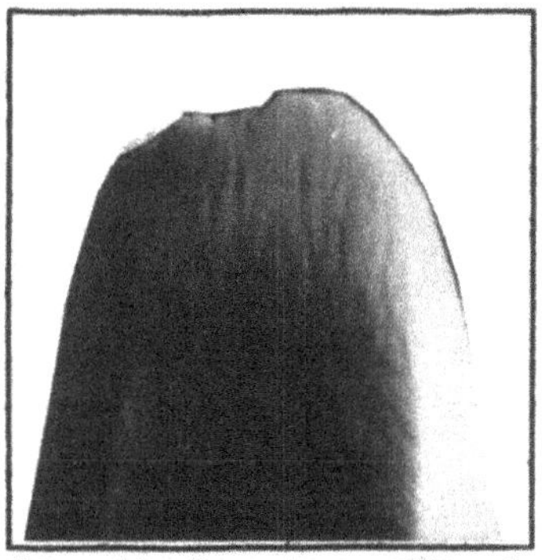

Plate 38: ExperimentalWood-working Chisel (2X)

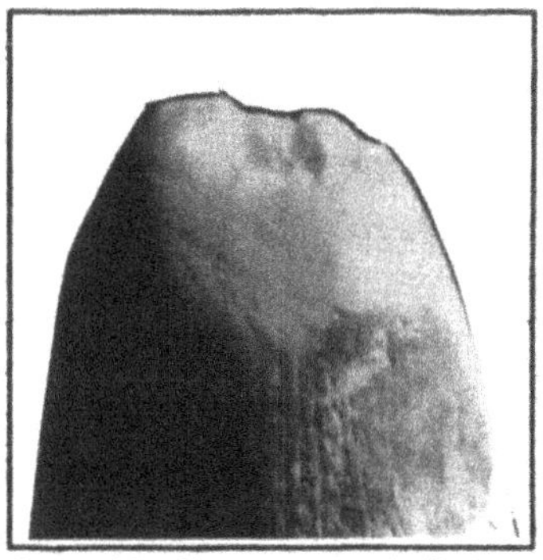

Plate 39: Experimental Chisel; Note Detached Chip

An attempt was then made to use the tool on the side of the wood block, in the manner of a chisel. The tool was held at about 45° to the surface of the wood, with the cancellous side of the bone down. About 1000 blows were struck with the bone mallet. The tool worked as a chisel, but it was not very efficient, allowing only a hole about 5 cm by 5 cm by 0.5 cm deep to be gouged. The following wear was observed. The blade at the tip became chipped, with the chips removed from the outer, upper surface of the tip. At the base the outer edge of the tool had become chipped due to the angle at which the mallet had struck the base. No very clear scratches were observed on this implement.

Plate 40: Experimental Chisel: Battering on Base

Another experimental tool was made from a *Bos* proximal metatarsal shaft with attached epiphysis. This tool was 97 mm long and 38 mm by 43 mm wide at the base. The shaft was split diagonally with a hammer blow, and a sharpened spatulate tip was ground into the narrow end of the shaft. The resultant blade, at a right angle to the shaft axis, was approximately 9 mm wide.

This tool was used as a gouge to remove chips of approximately 2 cm length from a willow stick. The stick was held in one hand while the tool was held by the butt end in the other hand. There was a strong tendency to increase the pressure on the tip by pressing with the thumb. The tool had a slight tendency to slip sideways on the wood. Otherwise it proved quite efficient, with little apparent dulling after about 200 chips. The tool was held with the cancellous side to the wood and not reversed during the course of the experiment.

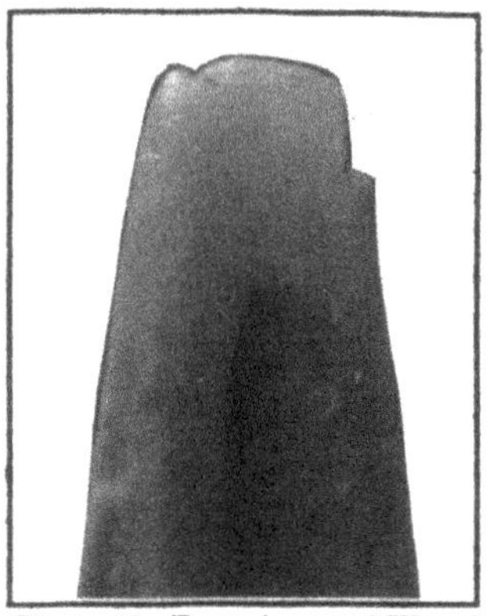

Plate 41: Experimental Wood Gouge (2X)

The following wear was observed on the tool. After 200 chips had been removed there was a slight polish apparent on the side of the tip facing the wood and a very slight rounding of the edge. After 500 chips this polish had increased and some of the marks of manufacture at the tip were beginning to wear away. The tip edge had become duller, and it was more difficult to penetrate the wood. After 1000 chips had been removed the polish and rounding were more marked, but the sharpness of the tip edge seemed to have stabilized. After 1400 chips a small flake was detached from the inner surface of the tip, evidently caused by prying a wood chip from the stick. After 2000 chips had been removed the marks of manufacture had all but been obliterated at the tip. The tip was polished to about 0.5 cm from the tip on both the inner and outer faces of the tip. The polish on the outer face was evidently due to friction from contact with the chip as it curls away from the wood. There were very fine striations or scratches parallel with the axis of the tip visible under the microscope at c. 50 diameters. Very little hand polish was visible on the base.

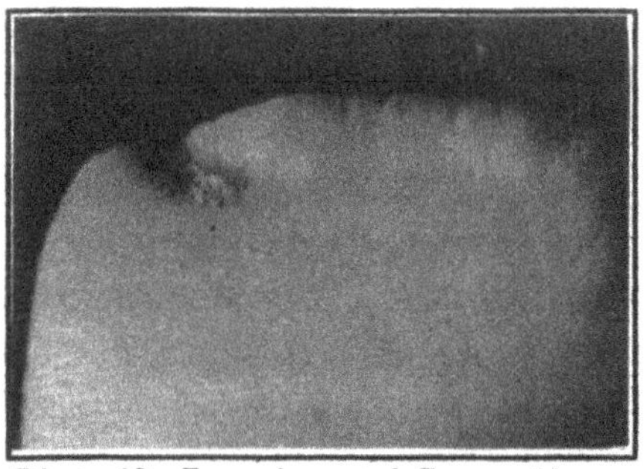

Plate 42: Experimental Gouge Above: (c. 6X) Moderate Rounding, Chipping, and Polish; Irregular Axial Scratches

Distinguishing between marks of manufacture and marks due to tool use

It is important in this kind of study to distinguish those marks which remain on a bone implement's surface as vestiges of the manufacture process from those marks which are the result of use-wear. Although in certain specific instances such a distinction can be quite difficult to make, in the majority of cases the source of surface marking can be determined with some confidence.

Experimental manufacture of bone tools, both by shaving with a flint implement and by abrasive techniques, has provided comparative specimens which illustrate the forms of surface markings characteristically left after manufacture. The reader should refer to the sections above for discussion of these markings.

In addition, several other criteria may be noted which may aid in distinguishing manufacture marks from use-wear.

1. It is unlikely that a tool will be used to work a material substantially harder than itself. There are, of course, obvious exceptions to this rule, the

most noteworthy of which is the use of a bone implement to retouch a flint edge. Generally, however, the use of a relatively soft tool to work a hard material would result in very rapid tool wear and very inefficient working of the material. In practice, clear scratches and gouges covering the surface of a bone implement are almost surely the result of manufacture if the following criteria are met.

a. If the worked material is softer than the bone, clear scratches and gouges resulting from tool-use can only have been caused by hard inclusions such as sand and dirt in the worked material. Since the inclusion of such material is random and irregular, scratches on the bone surface from such a source would be sparse. Although the alignment of such scratches may be patterned due to tool movement during use they would not be likely to cover the majority of the tool surface nor would they appear on every specimen of a given tool form.

b. If the direction of movement of the bone tool, as indicated by the alignment of the scratches, was convenient and practical for intentionally shaping the bone tool, then such scratches are likely to be the vestiges of manufacture. Comparison with experimentally manufactured bone tools and the experience gained in their manufacture aid in making this judgement.

2. It is more difficult to distinguish intentional polish from polish due to tool-use, and it is probable that there is no way to distinguish intentional polish from use-wear on the basis of the specific character of the polish in many cases. Intentional polish is likely to be distributed over the surface of an implement. In this study if polish has been found located only on protrusions and high areas, such as on the ends of long bones which would have been appropriate for use as handles, this polish has been considered as likely due to tool use. Similarly, polish confined to the very tips of sharpened implements is unlikely to be intentional and has been considered the effect of tool use.

An additional observation may be made which is important where both marks of manufacture and use-wear are present. Unless an implement is reworked or sharpened (and this is usually evident) marks due to tool-use will superimposed upon marks of manufacture. The effect of wear by the relatively soft worked material upon the bone surface will be to round away and partially or totally obliterate the marks of manufacture, leaving a more-or-less smooth surface. This effect is uneven, with only partial obliteration of manufacture marks visible on the majority of specimens. Scratches due to tool-use will then be superimposed on this relatively smooth surface, appearing sharp-edged and relatively unworn. This effect occurs on the majority of worn specimens, leaving no difficulty in distinguishing the two forms of surface marking.

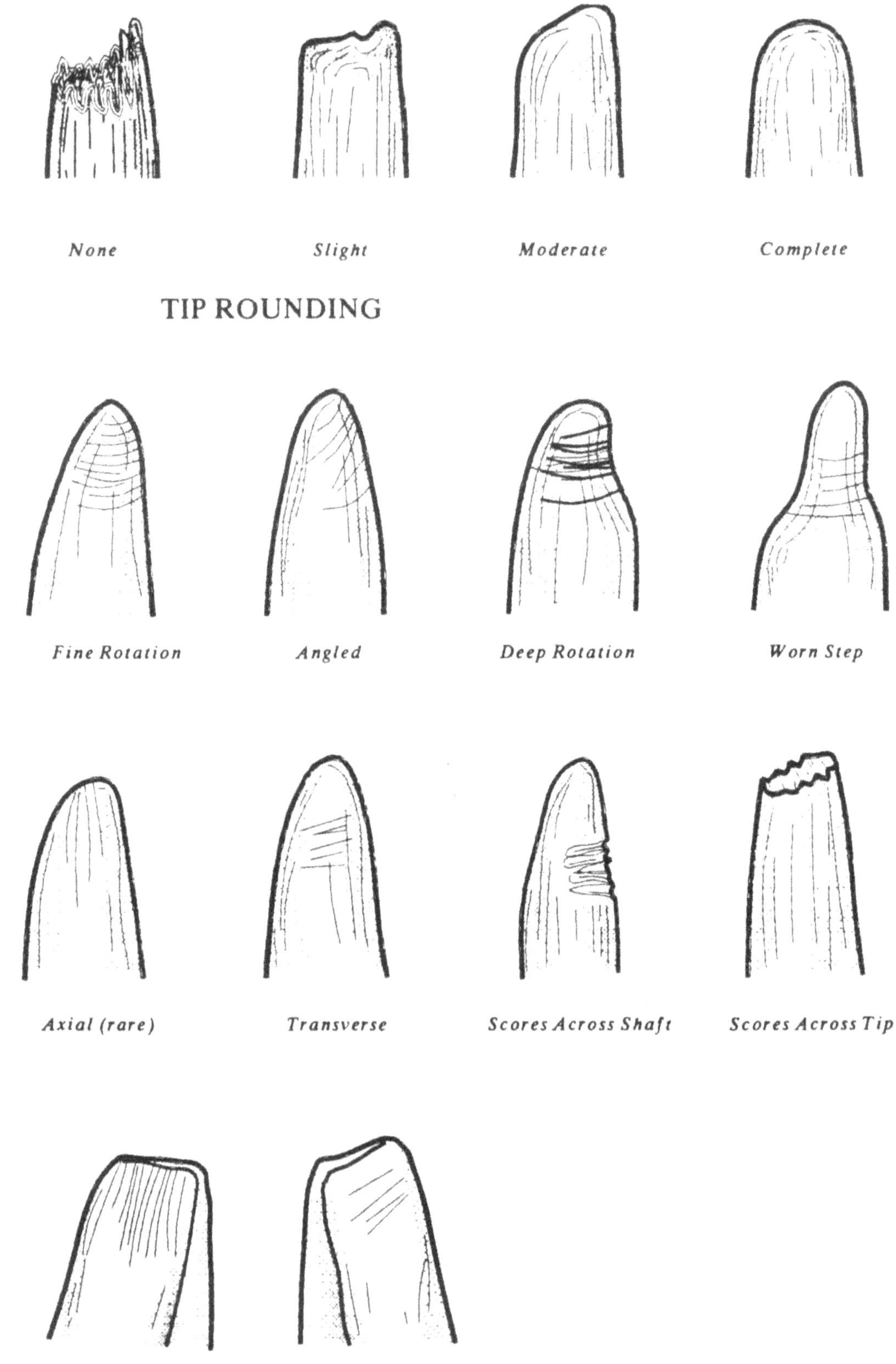

Figure 5: Scratch and Score Wear Patterns

TABLE 1

CRITERIA FOR DISTINGUISHING MARKS DUE TO MANUFACTURE PROCESSES FROM MARKS DUE TO TOOL USE

Surface Marks	Manufacture Marks	Marks Due to Tool-Use
Scratches and Gouges	Uniformly distributed over surface.	Sparsely distributed; does not necessarily appear on every specimen of a given form.
Scratches and Gouges	Surface examination indicates scratches are the result of cutting with a sharp edge or abrasion against a hard, rough material. Scratches are the result of the cutting of individual sharp grains.	
Scratches and Gouges	Direction of scratch indicates relative motions which would be convenient and practical for the shaping of a given tool.	Direction of scratches indicates relative motions which would be practical for tool-use but unlikely in implement manufacture, or randomly-oriented motions (which are most likely due to carrying, storage, or post-depositional action).
Scratches and Gouges	Often partially worn away by subsequent friction against a soft material (use).	Often overlie polish due to friction with a soft material, and usually sharp-edged and distinct.
Polish	Likely to be uniformly distributed over the tool surface.Confined to raised areas and to sharpened tips and edges.	

CHAPTER VI

The Functions of Natufian Bone Tools

Points and Pointed Implements

Many Natufian and Zagros Protoneolithic pointed implements probably served as perforators for various materials. As illustrated in the previous chapter, the possibility of such a use is to a great degree dependant upon the tool possessing a sufficiently small tip size to permit penetration of the worked material.

Figures 6 and 7 illustrate the distribution of tip diameters of the Natufian and Zagros Protoneolithic points and pointed implements. This dimension was carefully measured in all these specimens. Its distribution was plotted in order to uncover any bimodal or polymodal curve which would suggest

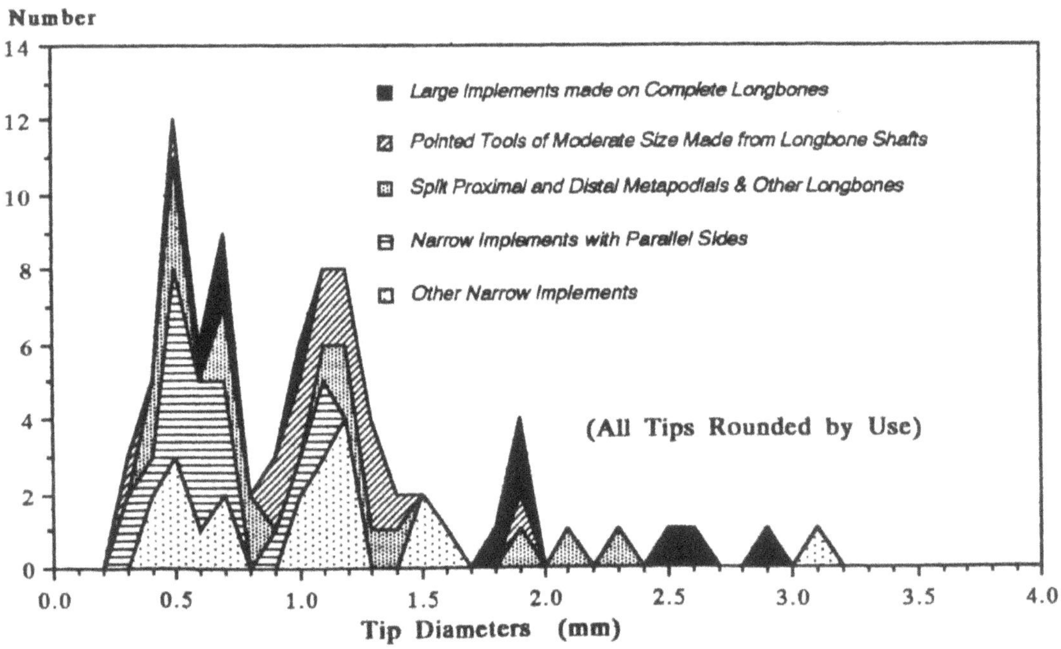

Figure 6: Natufian Pointed Implements: Tip Diameters

distinct uses for pointed implements with markedly different tip diameters. In the Zagros Protoneolithic case there is no clear polymodality. Among the Natufian pointed implements there are two peaks, centered upon 0.5 mm and 1.1 mm, with few tip diameters falling at 0.8 mm. There is no clear correlation, however, between these peaks and the various tool forms. The principal exception is parallel sided implements, which have, on the average, smaller tips than the other tools (See Figure 8). It is likely that the two peaks represent, respectively, points with unbroken tips and points on which the very tip has broken off, either in fabrication or use.

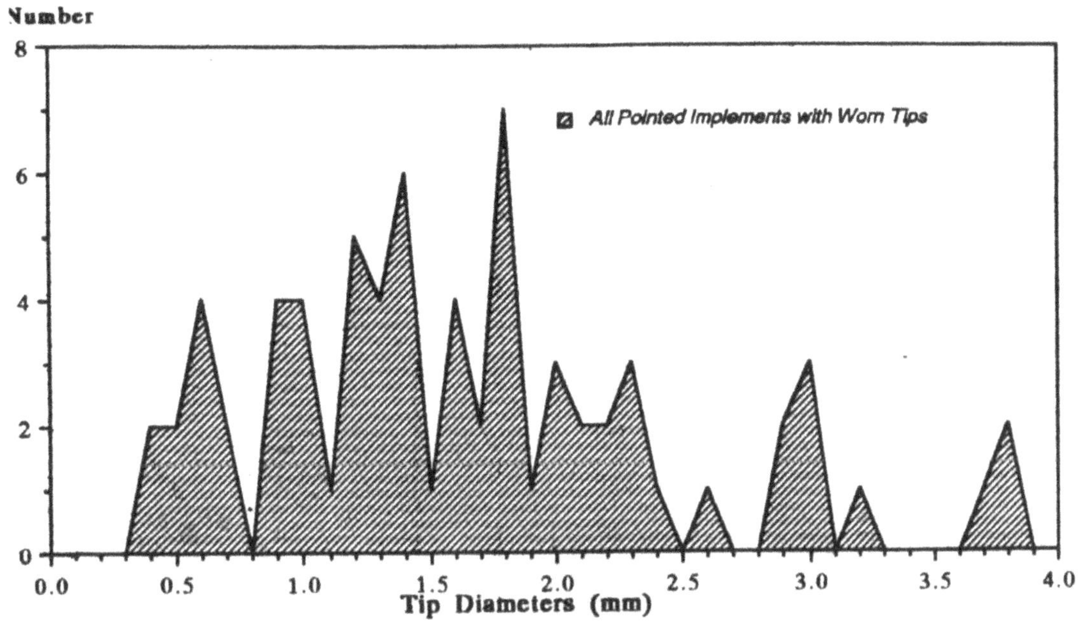

Figure 7: Zagros Protoneolithic Pointed Implements: Tip Diameters

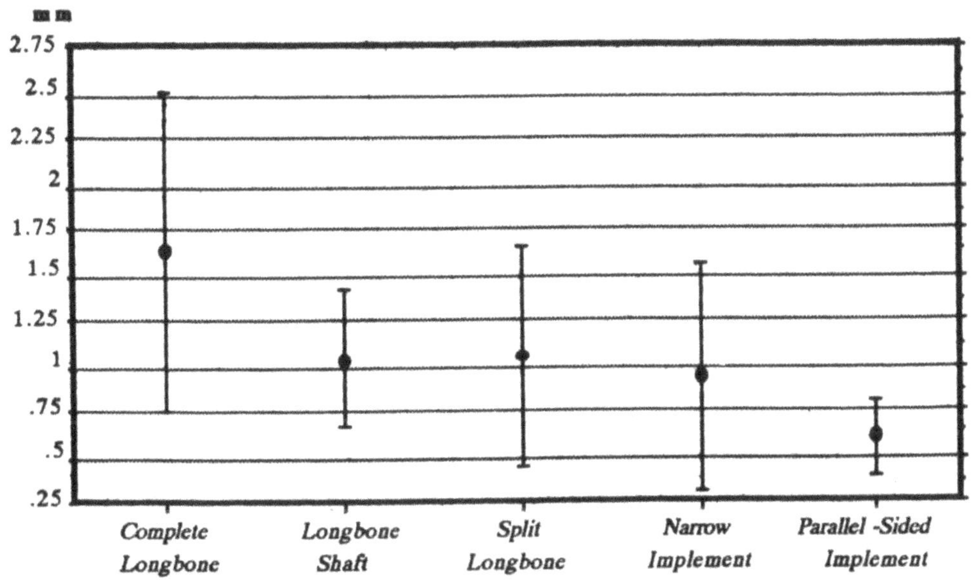

Figure 8: Tip Diameters of Natufian Pointed Implement; Bars Represent 1 Standard Deviation

Observed wear-patterns on points and pointed implements

In order to formulate specific hypotheses as to the use of these tools it will be necessary to examine in detail the wear appearing on them. The first group to be examined are all quite large; nine tools made on complete distal metapodials, four on other large bones, and two on large bone shaft fragments (total 17).

The nine implements on distal metapodials are quite uniform in shape, varying primarily in length. Eight of these are made on mature bones, while one is missing the distal epiphysis. The two tools on distal tibias and one of the tools on a proximal metapodial are similar. The two implements on bone fragments and those made on a large ulna are different in shape but similar in overall size. The remaining complete implement from Nahal Oren is unique in that while the tip itself is made on a gazelle proximal metatarsal, the tarsals, astragalus, and calcaneus remain attached. This implement must have been used with much or all of the connecting tendon still in place.

All these specimens show some degree of wear indicating that they were handled more or less extensively. The protruding high areas of the epiphysis or basal end of the tool is usually slightly rounded and shows some degree of polish which could only be the result of abrasion against a soft surface (almost surely the hand of the user). Under the microscope these areas show a smooth, satiny texture lacking any fine scratches to indicate a preferred direction of abrasion. Such polish cannot be attributed to post-depositional chemical effects as it does not occur in the recesses of the bone. The most marked rounding and polish is found surrounding the shaft of the tool about midway between the tip and the basal end. This polish is clearly of mechanical origin, appearing only on high areas - those areas which would be directly contacted while holding the implement firmly in the hand. The area surrounding the basal end, where the bone surface is concave, is usually free from polish as is the area of the shaft immediately adjacent to the tip. The area of polish, is not sharply bordered but rather grades away smoothly to the natural or worked surface of the bone. Examined under moderate magnification (60-100 diameters) the polished surface usually shows numerous fine scratches superimposed on the polish. These scratches rarely show any clear ordering or pattern but tend to be randomly oriented. The scratches probably resulted from a combination of causes: abrasion from the soil on the hand of the user, random abrasions during the period of tool use, and further post-depositional abrasions.

Plate 43: HAY 135 (¼X)

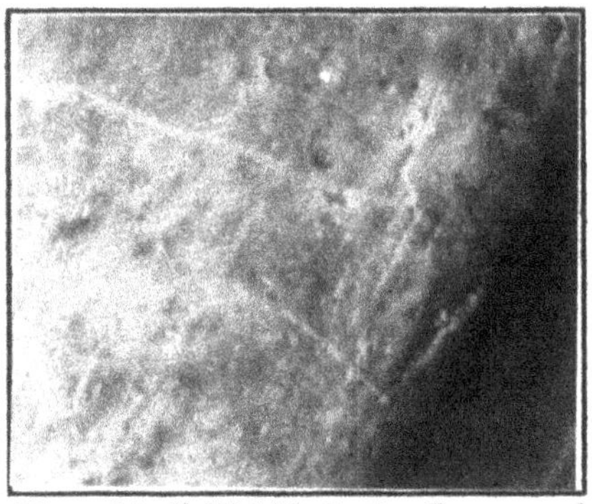

Plate 44: HAY 135: Satiny Texture of Area Smoothed by Hand Polish

More important is the depth of polish and rounding attained by some of the tools. On some specimens the natural bone surface and many of the marks of manufacture have been worn away by handling, suggesting a very protracted period of use. On one specimen from Kebarah Cave (KEB 19) the shaft and basal end show substantial rounding and polish from handling, nearly obliterating the marks of manufacture. The very tip, however, while complete appears completely unworn and the marks of manufacture are clearly visible for approximately 1 cm from the tip. There is no rounding, polish, or scratches. This discrepancy of wear clearly indicates that this tool has been resharpened.

As it was necessary to maintain a narrow tip on these implements it is evident that they were used to penetrate a material, fabric, or closely-spaced structure of some kind. If it is assumed that these similar tools served similar purposes, the following observations may be made:

1. Complete tips range from 0.6 to 2.9 mm in diameter[14]. The entire range of sizes, including the largest, shows some degree of rounding of the tip edges. Only two of the specimens have no rounding of the tip: the apparently resharpened implement from Kebarah and a tool from Hayonim Cave with a heavily chipped tip. Seven of these specimens have missing tips, but two of these, one 1.8 mm and one 2.5 mm in diameter, show slight rounding of the break edge indicating continued use after breakage. The four specimens with broken tip 3.0 mm in diameter or greater are all unworn. This evidence suggests that the maximum useful tip diameter was about 3 mm for these implements.

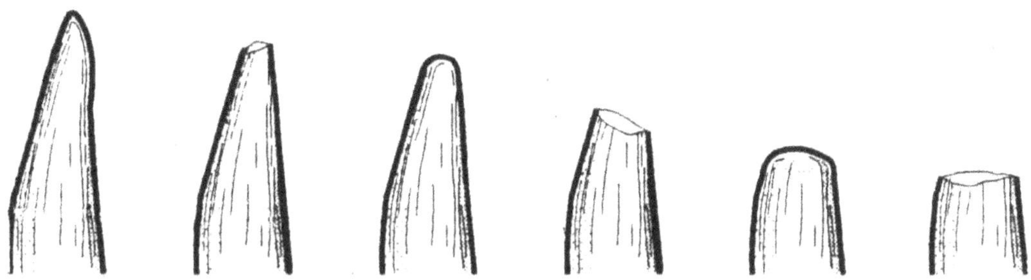

Figure 9: Proposed Wear-History of a Bone Tip

2. Most of the larger tips designated as "complete" show a chip or chips missing around the tip edges. The chips appear to have originated from sideways pressure on the tip. Such pressure must also have resulted in the breakage of many tips. Only one specimen with a tip diameter greater than 1.5 mm is not chipped or broken. It is likely that most tips were originally small in diameter but that this diameter increased with chipping and breakage until a diameter was reached at which the tool was no longer usable. It was then either resharpened or discarded.

---

[14] A full listing of these measurements, plus a complete description of all wear-patterns, may be found in Campana, 1980.

3. With few exceptions these implements show an area of polish extending up the shaft from the tip, usually for about 1 cm, although the boundaries are indistinct. This polish clearly indicates the functional area of the tool. At this distance from the tip the width of the tool shaft averages about 5 mm. Therefore the use of these tools for making fine perforations in leather or hide is precluded. Only holes of a very large size might have been made.

4. The most important source of evidence is the pattern of fine scratches often appearing near the tip. Of the seventeen complete tools, four specimens show heavy scratches due to tool rotation scratches surrounding the tip, four specimens show fine rotation scratches, on has scratches leading back from the tip in an angled spiral, and two have patches of transverse scratches across the tip.

Plate 45: KEB 77 (½X)

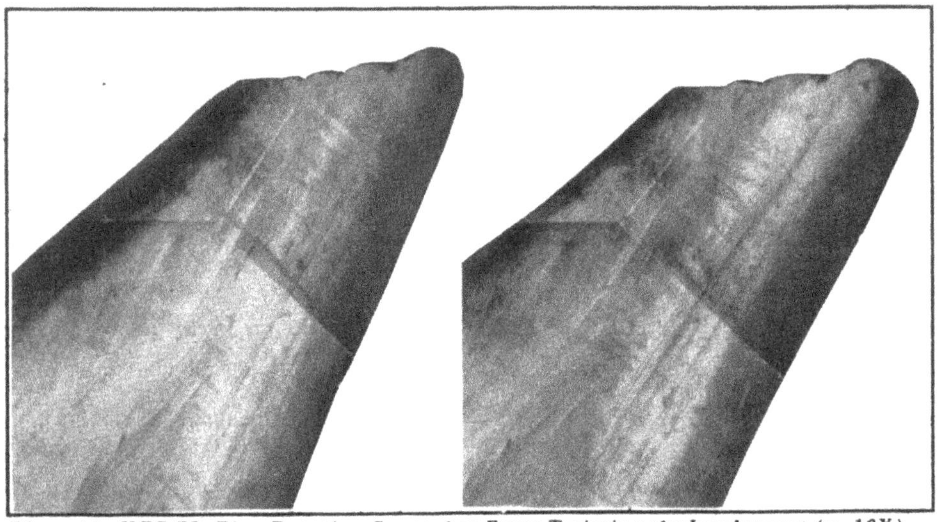

Plate 46: KEB 77; Fine Rotation Scratches From Twisting the Implement (c. 10X)

As eight of the specimens (47%) were clearly twisted in the course of use this may be interpreted as a normal manner of use. A further clue to tool function may be found in the position of these scratches due to tool rotation. Such scratches are found at a minimum of 6 mm to a maximum of 34 mm from the tip. The diameter of the tool at the scratch position ranges from 1.9 to 12 mm.

Conclusions: On the basis of the above evidence it is relatively easy to characterize the mode of use of these implements. Held in the hand, they were pushed into the worked material. Often the tool was twisted to ease the penetration. The high proportion of specimens with deep rotation scratches as well as the number of specimens with heavily chipped tips indicates considerable force was used. Certain possible uses, as "picks" or "daggers" may be eliminated, as such used would not lead to rotation scratches. It is far more difficult to identify the worked material with certainty. About 80% of the specimens could have effectively penetrated hides but the remainder had tips which were probably too large. On the whole, the large diameter of the shaft at maximum penetration into the worked material indicates that these tools were used to make rather large openings. If they were used in hide working, some of the tools with large tips could only have been used to enlarge holes made by some other implement. In view of the high proportion of tools with deep scratches it is perhaps more likely that these tools were used in the manipulation of more abrasive materials such as heavy basketry or matting than in the working of hides.

Natufian implements:
Split proximal metapodials and split distal metapodials

Eighteen Natufian implements are made from split proximal metapodials and nine from split distal metapodials. Most of these specimens show marked rounding and polishing of the high points of the epiphysis and around the central area of the shaft indicating extensive handling. Four specimens (distal metapodials) have had perforations made through the distal epiphysis. These holes all show some degree of rounding and polish from abrasion within them and in three cases this is more marked on the edge of the perforation adjacent to the base. This wear is surely due to the rubbing of a cord, as it closely matches wear obtained by experiment. It seems most likely that this cord would have been tied through the hole and used for suspension of the implement. Alternatively, the tool might have been used to draw a cord through a hide or fabric in the manner of a bodkin. For three of the specimens this is unlikely on several counts: (1) the perforations are not so worn as might be expected had a cord been pulled through them consistently, (2) the epiphysis at the base is too expanded for easy passage through a fabric, webbing, or netting as might be expected of a bodkin, (3) the remaining six specimens, otherwise similar, do not have a perforation. For the fourth perforated specimen, however, these arguments cannot be made so forcefully as the perforation is more markedly worn and the base is reduced in width so that it might have passed through a weft easily. The implement is also slightly polished overall.

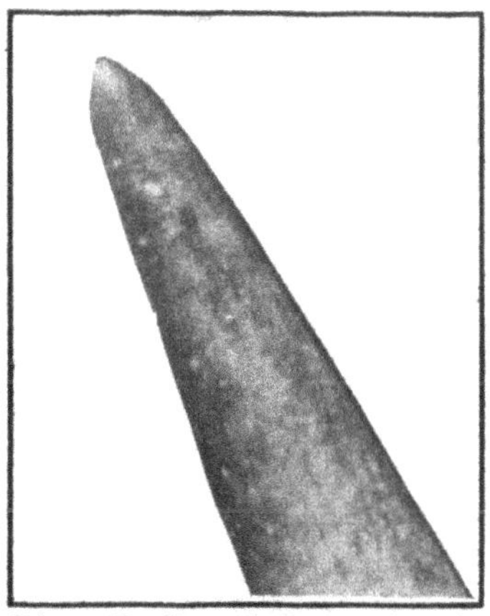

Plate 47: Rounded, completetip of split proximal metapodial tool HAY 156 (10X)

Plate 48: HAY 156 (½X)

For these two groups of tools the following observations are made:

1. For split proximal metapodials the diameter of the eight complete tips ranges from 0.4 to 1.4 mm. Of the ten broken tips four show wear from continued use; the largest of these is 2.3 mm in diameter. For split distal metapodials the diameter of the five complete tips ranges from 0.4 to 1.4 mm as well. None of the four broken specimens shows wear.

2. Many of the specimens show patterned scratches at the tip. Five of the split proximal metapodial tools bear angled scratches at the very tip (or at the tip break). The spiral path of these scratches indicates that the tool was slightly twisted as it was pressed into the worked material. For split proximal metapodials the maximum distance to any light rotation scratches present ranges from 3.7 mm to 30 mm. The shaft diameter at these scratches ranges from 1.9 mm to 9 mm. The maximum shaft diameter at the scratches could be determined for only one of the split distal metapodials; this equalled 8 mm diameter at 15 mm from the tip.

Plate 49: WAD 7 (½X)

Plate 50: KEB 20 (½X)

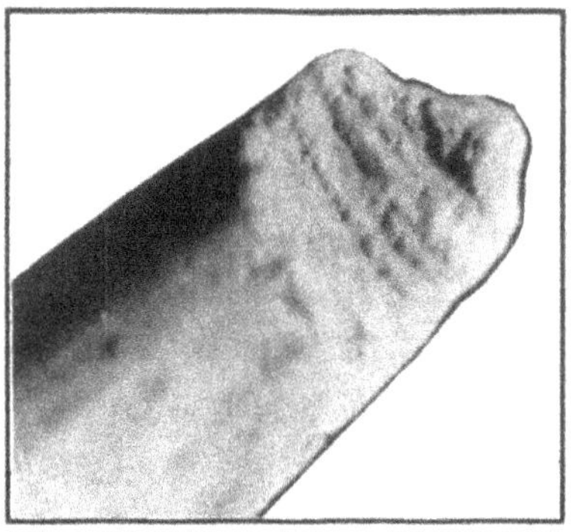

Plate 51: Scores Across Tip of Split Proximal Metapodial Tool WAD 7 (10X)

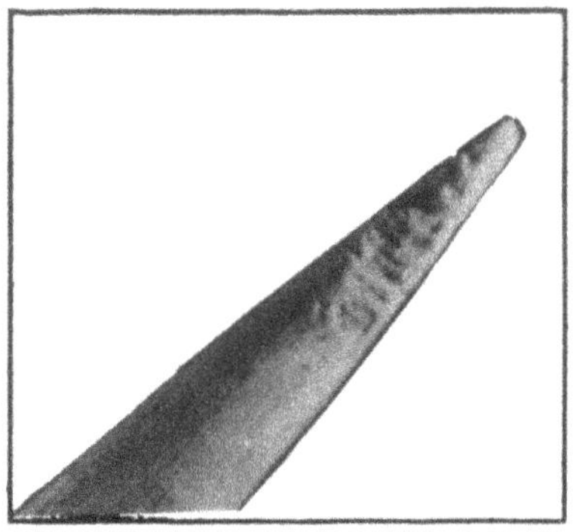

Plate 52: Scores Across Tip of Split Distal Metapodial Tool KEB 20 (2X)

3. Three specimens on split proximal metapodials show deep scores running across the tip, the shaft very near the tip, or both. Two specimens on split distal metapodials show similar scores.

There are three additional similar but slightly smaller specimens made on other long bones. Two tools from Hayonim Cave are rounded and polished over most of their surfaces indicating extensive handling. Both show smoothing of the shaft to several millimeters from the tip. Neither shows any clear scratches. The third tool from El Wad shows very little wear or polish of any kind. Twenty point fragments appear to have been broken from implements of similar type. Their tip diameters range from 0.3 mm to 1.4 mm on the unbroken specimens. One broken tip 2.6 mm in diameter and two 1.3 mm in diameter show signs of continued use. Rotation scratches appear on 50% of these specimens.

TABLE 2

NATUFIAN POINTED IMPLEMENTS: SCRATCH PATTERNS

Scratches	Split Distal Metapodials (N = 9)	Split Prox. Metapodials (N = 18)	Fragments (N = 20)
Total Rotation	3 (33.3%)	10 (55.6%)	10 (50.0%)
Light Rotation	3 (33.3%)	5 (27.8%)	6 (30.0%)
Heavy Rotation	0	0	3 (15.0%)
Transverse	1 (11.1%)	1 (5.6%)	0
Angled at Tip	0	5 (27.8%)	1 (5.0%)

*Some implements bear no wear; others more show more than one pattern.

Conclusions: The wear appearing on these implements indicates that the manner of their use was generally similar to the larger implements discussed above. However, the following differences may be observed: (1) the tip diameters are somewhat smaller; (2) the proportion of angled scratches at the tip is larger; (3) a small proportion of these tools have scores across the tip, which do not occur on the larger implements. On one specimen scores across the tip are found associated with rotation scratches on the shaft. It is difficult to reconcile these very different forms of wear. Scores on these tools are too common to be dismissed as accidental but are to rare to considered representative of normal use. Scores are found on other pointed implements as well. Their significance will be discussed below.

The occurrence of angled scratches strongly suggests that these implements were used as perforators, probably for hides. However, the diameter of the shaft at the rotation scratches is nearly the same as on the larger implements. These tools were probably not used for fine work.

There is considerable similarity of form between tools made on complete bones and those made on split bones. As these tools differ primarily in size it seems probable that there would be an overlap in the functions to which they were put. The manner of use of all these tools is similar: insertion and twisting of the tip in order to create a hole or opening. The material being worked is not easily identified, but may best be seen in terms of a continuum of probability: tools on split metapodials were most likely used for perforating hides while large implements may have been used with more open-textured materials such as matting or basketry, but many tools could have been used for either (or both) purpose.

Natufian implements:
Narrow and parallel-sided pointed implements

The Natufian bone tool assemblages included seven complete implements and 18 fragments with narrow, straight shafts. The shafts of these tools have parallel faces and sides. As the fragments can fairly securely be assigned to this group all the specimens will be considered together. One of these specimens has a very wide shaft with a perforation at its base; this artifact will be considered separately. The width of the remaining specimens is fairly uniformly distributed from a minimum of 3.0 mm to a maximum of 9.3 mm. Thickness ranges from 2.9 mm to 4.9 mm. The narrowest tools are nearly round in cross-section while the larger tools are quite broad. This may be expected given the limiting thickness of the bone shaft. Nine specimens have complete tips. These range in diameter from 0.3 mm to 1.0 mm. One of the wider implements (SHK 2) has a small perforation at its base. In this respect it is similar to the very broad specimen WAD 9. Another, HAY 152, has an indentation at its base which appears to be an incomplete perforation.

Another 18 implements have generally similar dimensions but the sides of the shaft range from being somewhat curved to highly irregular in shape. Fifteen of the points appear also to have belonged to this group or to the group with parallel sides but are too short to be placed with confidence. The group as a whole ranges in width from 4.3 mm to 17.0 mm and in thickness from 2.4 mm to 9.0 mm. Of these tools SHK 14 has a perforation at the base and another, WAD 13, has an incomplete perforation. One implement is unusual in that its tip is ovate rather than round (0.9 mm by 0.5 mm). This specimen bears a unique wear-pattern and will be discussed in detail below.

It is probable that the implements with parallel sides and the narrower specimens above could have served similar functions. The largest implements, with broad tips, are likely to have served a different function, but it is difficult to fix a dividing line between the potential uses on form or size alone.

These tools are all polished and rounded to some degree, and this wear (with some exceptions) is fairly uniformly distributed over the surface. Those specimens which are designated as complete show distinct rounding of the break at the basal end. It is possible that some of those specimens which have been considered fragments were in fact complete, but as there was no visible rounding at the basal break this could not be ascertained. The degree of rounding and smoothing varies somewhat on these specimens, being more marked on the parallel portion of the shaft, the base, and on the area immediately adjacent to the tip than on the conical portion of the lower shaft. There are usually randomly-oriented scratches visible in the polish of the upper shaft.

The following observations are made:

1. The diameter of the complete tips of the parallel-sided tools is quite small (mean = 0.56 mm). Although 15 specimens have broken tips only four show signs of continued use; the largest of these is 1.0 mm in diameter. The seven additional possible parallel-sided fragments have tip diameters ranging from 0.25 mm to 1.0 mm. Only one tip is broken, and it (0.7 mm) shows continued use. It may be noted that of the other narrow implements eight complete tools and twelve fragments have greatest widths which are smaller than the broadest of the parallel-sided tools (9.3 mm) and are therefore roughly comparable. The diameters of the twelve complete tips range from 0.4 mm to 1.45 mm. Four broken tips show continued use; these range from 0.9 mm to 1.5 mm. For all the above specimens this evidence is consistent with the use of these tools primarily as perforators.

With the exception of KEB 5, which will be discussed separately, there are in addition nine complete and three fragmentary specimens with curved or irregular sides wider than 9.3 mm. Six have unbroken tips ranging from 0.45 mm to 1.0 mm. Two tips show continued use after breakage; the larger of these is 3.1 mm in diameter.

2. Patterned scratches at the tip are distributed as shown in Table 3. On the parallel-sided fragments the position of the rotation scratches ranges from 4.0 mm to 48.8 mm from the tip. On two broken fragments these scratches are found over the entire length of the shaft to the basal break, 36.0 and 48.8 mm from the tip. The diameter of the shaft at these scratches ranges from 2.0 mm to 8.7 mm. For the curved-sided specimens rotation scratches are found 2.3 mm to 70.0 mm from the tip, and the shaft ranges from 1.7 mm to 6.0 mm in diameter at this point. In the case of HAY 199 such scratches are seen to extend nearly two-thirds of the length of the complete implement back from the tip.

3. Four parallel-sides and one curved-sided specimen show scores across the shaft adjacent to the tip. As with the previous tool group these scores probably do not represent a primary tool use. They will be discussed separately below.

Included in the parallel-sided group is a specimen with a perforation near the base (SHK 2) and a similar specimen with an incomplete perforation. The complete perforation is 3.1 mm in diameter and shows rounding and polish on the edge nearest the base, indicating that a cord passed through it. This implement shows transverse rather than rotation scratches across the tip and also across the central area of the shaft. SHK 14, with curved sides, is very similar, but shows angled scratches at the tip as well as many transverse scratches across the shaft. The perforation at the base, 2.1 mm in diameter, is very well rounded and polished, perhaps slightly more on the basal edge. WAD 9 is somewhat similar to these specimens but is considerably wider (15.0 mm). The perforation at the base of WAD 9 is 3.4 mm in diameter and heavily rounded on the basal edge. It has a small, round tip (0.9 mm) with light rotation scratches surrounding it to a distance of 10 mm from the tip (shaft diameter 10 mm at that point). Transverse scratches also run across the shaft.

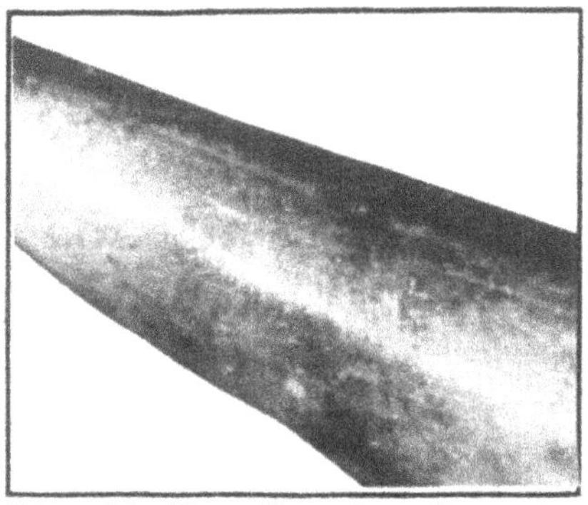

Plate 54: Fine Rotation Scratches Surround Tip of WAD 9 (10X)

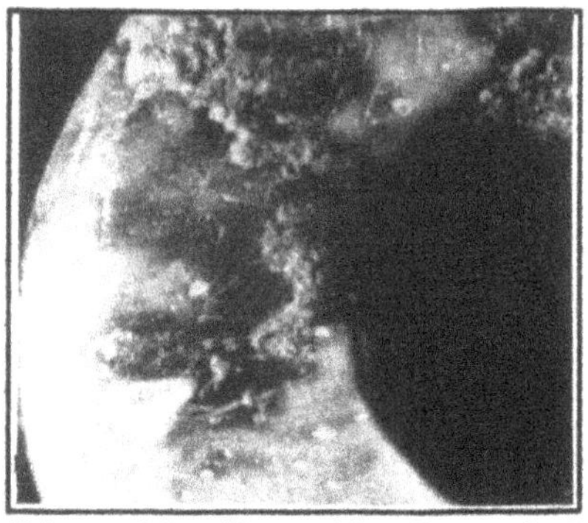

Plate 55: Heavily Rounded Perforation of WAD 9 (10X)

Plate 53: WAD 9 (½X)

Several other specimens require additional comment. One fragment (KEB 68) is very densely surrounded by fine rotation scratches. A constriction around the tip may have been worn into the bone by this rotary action. Another fragment (KEB 69) conforms in general shape to this group but as it is incomplete it could possibly have been a double-pointed implement instead. It will described separately as it has several unique features. The fragment KEB 62 appears to have been worked from a fragment of a tooth, the enamel being visible on one side. It is somewhat asymmetrical and shows very little wear. The complete implement KEB 5, unique in terms of wear, will be discussed in detail separately.

Conclusions: It is clear that in normal use the majority of these tools were pressed into and through the worked material, often with a twisting motion to assist penetration. The complete points are usually very small, and larger broken points which shows signs of continued use are rare. This evidence points to the principal use of these tools as perforators of a compact material such as hide. Several specimens must have been pushed a large fraction of their length through this material. Most of the parallel-sided tools and many of the curved-sided tools could have been pushed completely through if desired, but their is little indication that this was done. The wear on the upper shafts and bases appears more likely to have resulted from handling, and rotation scratches do not appear near the bases of the incomplete implements. Indeed there is little purpose in forcing these implements all the way through as the majority are eyeless and could not have functioned in the manner of modern sewing needles. It is more likely that they were used to make holes for later lacing.

TABLE 3

NATUFIAN IMPLEMENTS AND POINTS ON FRAGMENTS: SCRATCH PATTERNS

Scratches	Parallel Sides (N = 24)	Curved or irregular sides *Narrow* (N = 20)	*Wide* (N = 12)
Total rotation	12 (50.0%)	8 (40.0%)	6 (50.0%)
Light Rotation	9 (37.5%)	5 (25.0%)	4 (33.3%)
Heavy rotation	1 (4.2%)	1 (5.0%)	1 (8.3%)
Angled at tip	2 (8.3%)	2 (10.0%)	1 (8.3%)
Transverse	1 (4.2%)	1 (5.0%)	0 (0.0%)

	Narrow specimens All* (N = 50)	Curved/irregular sides; All (N = 32)	Overall (N = 63)
Total rotation	25 (50.0%)	14 (43.8%)	31 (49.2%)
Light Rotation	15 (30.0%)	9 (28.1%)	19 (30.2%)
Heavy rotation	4 (8.0%)	2 (6.3%)	5 (7.9%)
Angled at tip	6 (12.0%)	3 (9.4%)	7 (11.1%)
Transverse	2 (4.0%)	1 (3.1%)	2 (3.2%)

*Parallel, curved, and possible parallel sides.

Those implements with perforated bases present more difficulty. Potentially these tools could have been used simply as perforators with a suspension cord, or they many have been used in the manner of bodkins. The area of the base opposite the worn area of the perforation was examined for rounding or striations that might indicate that a cord had been tightly tied around for suspension, but no such evidence was found. This is by no means conclusive however. The wear-pattern on the hole is not dissimilar to that found on Native American matting needles. The perforation of WAD 9 also shows fine scratches passing through the hole on the basal edge. The tips of the three perforated implements are small (0.47 mm, 0.9 mm, 1.1 mm) and show angled, light rotation, and transverse scratches across the tip, respectively. Only SHK 14, with the smallest tip diameter, has a compete tip, which is well worn. The others show no wear at the tip break. It appears likely that these tools were used as perforators in the same way as the other specimens of similar shape. SHK 2 and SHK 14 are sufficiently narrow to have readily penetrated, carrying a lace in the hole; WAD 9, however, is 15 mm wide at its largest point, and at first does not seem likely for this use.

Plate 56: SHK 14 (½X)

Examination of the tip, however, shows clear rotation scratches up to a shaft diameter of 10 mm. As such a large hole must have been made it is therefore possible that the tool was passed all the way through and used for rather crude lacing. The sides of WAD 9 are considerably more polished and rounded than the faces. This might be the result of greater friction occurring at the sides as the tool was passed through the hole, but could alternatively have resulted from handling. No clear axial scratches are visible in this polish, as would provide conclusive evidence, but the difficulty of observing axial scratches has been noted. It would be possible for such tools to have been used in netting, matting, or basketry, but it would be difficult to reconcile such uses with the small, sharp points and the evidence for use as perforators.

The existence of two specimens with incomplete perforations further suggests that the functions of the perforated tools and the similarly-shaped unperforated ones were probably related.

Unique specimens:

Several unique specimens among the narrow pointed implements merit individual discussion.

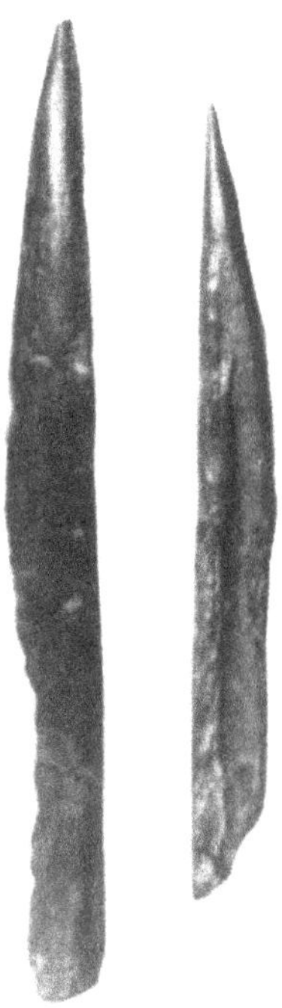

Plate 57: HAY 61 & HAY 25 (½X)

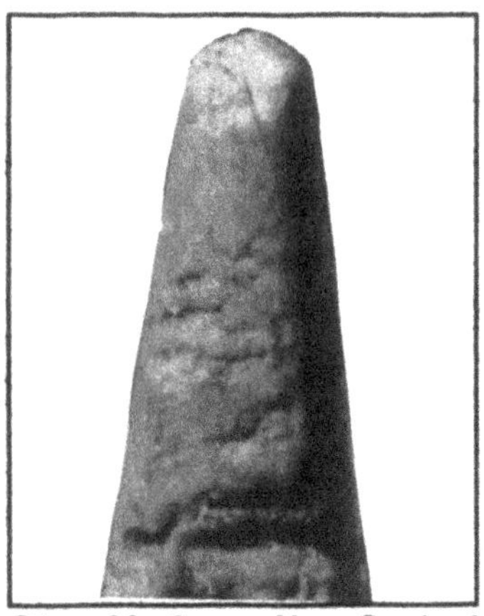

Plate 58: Scores Near Conical Tip of HAY 25 (10X)

Coarse Perforators: HAY 25 and HAY 61 are similarly large tools (135.5 mm and 180 mm long respectively). Both are made on large long bone shaft fragments and are slightly broader at the center portion of the shaft than at the base. This central area shows very marked rounding and polishing along the sides, indicating that the tool was probably gripped at this point rather than at the base. The tip of HAY 25 is 1.2 mm in diameter, but although it is complete, it is shaped as a very shallow cone with a marked shoulder at its periphery, rather than the more usual semi-spherical shape. It is well rounded and polished. The shaft is surrounded by many fine rotation scratches to about 15 mm back from the tip (the shaft is 5.8 mm wide at this point). There are several scores across the tip at 4 mm back. The tip of HAY 61, 2.1 mm in diameter, had been broken off and the edge of the break heavily chipped, but these edges are completely rounded and polished by continued use. The shaft is similarly surrounded by fine rotation scratches to about 20 mm from the tip (8.5 mm wide at that point). These tools were clearly twisted in use, but could not have been used to make fine perforations. Their function is likely to have been similar to the large implement described above.

Possible Hafted Implement: KEB 69 is quite similar in form to other parallel-sided fragments, but it is unusual in that one face is crossed by many nearly parallel transverse scores, which are clearly intentional. These scores run slightly around either side, but do not occur on the opposite face. The edges and protrusions of these scores are very well rounded and polished. The flat face opposite the scores is also rounded and polished, and the entire area is covered with very many, very clear, fairly fine parallel scratches running across it and around the sides. The tip is missing, but the break, 1.0 mm in diameter, is slightly rounded. A fine rotation scratch occurs 4 mm from the tip. The break at the base of this specimen is sharp, so it is not possible to determine whether this tool was a double-point; there is no narrowing, however, to suggest this. By far the most likely interpretation of this pattern of scores and scratches is that a cord was tightly secured around the shaft at this point, the scores serving to hold it in place. Possibly this was a form of hafting, but as the specimen is incomplete it is difficult to speculate further.

Plate 60: Slightly Rounded Tip of KEB 69 (10X)

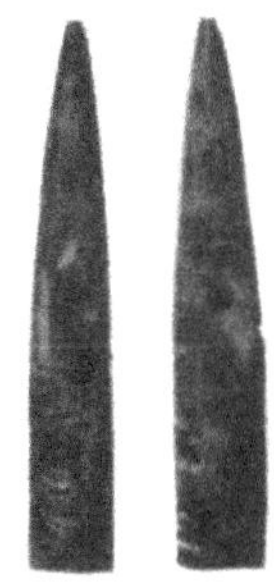

Plate 59: KEB 69 (1X)

Plate 61: Multiple Scores Across Face of KEB 69 (10X)

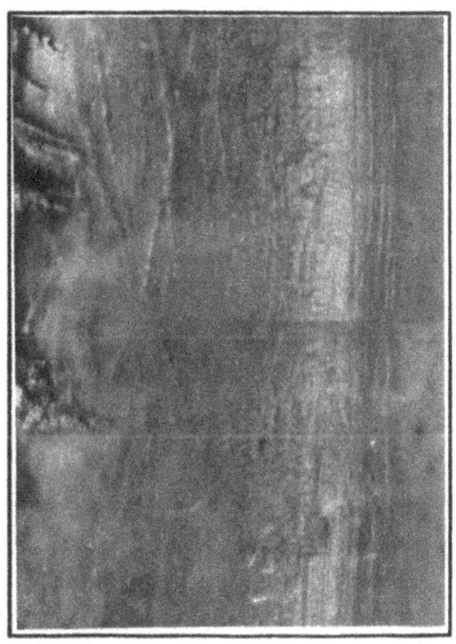

Plate 62: Multiple Fine Scratches Across Face of KEB 69 (10X)

Possible Basketry / Weaving Tool: KEB 5 is the most interesting specimen of this group. It is quite different from the others, although similar in overall form. This tool could not have been a simple perforator, as the

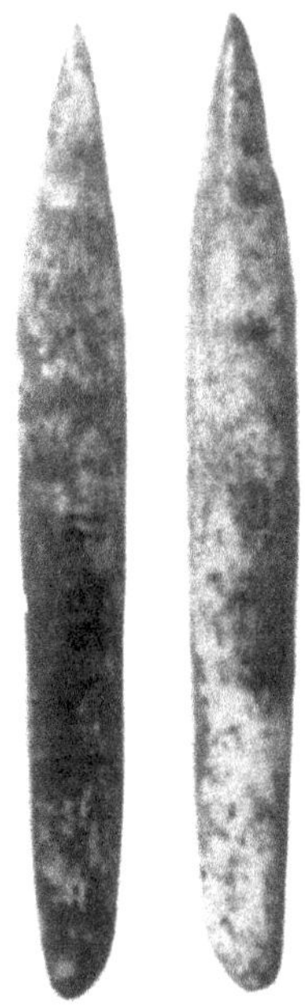

Plate 63: KEB 5 (1X)

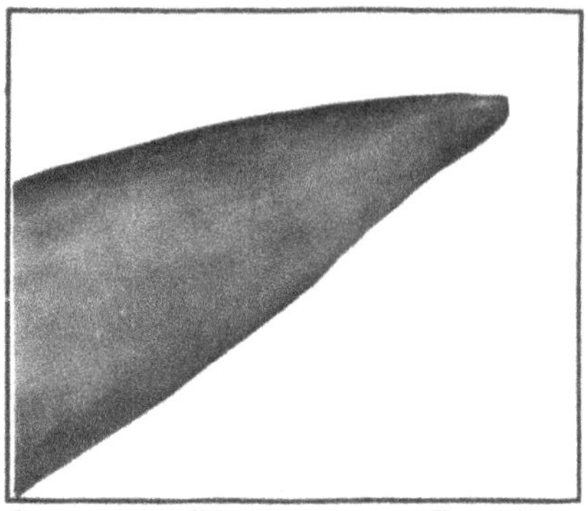

Plate 64: KEB 5: Flattened Tip; Edge View (10X)

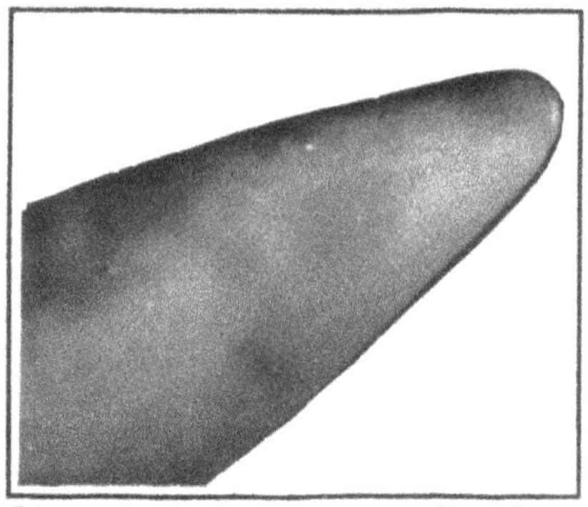

Plate 65: KEB 5: Flattened Tip, Face View (10X)

very tip, which is complete, is not round but ovoid, measuring 0.5 by 0.9 mm. The edge of the tip is completely rounded and polished. There are no rotation scratches on the implement, but there are very many parallel transverse scratches running across both the inner and outer faces to about 15 mm from the tip. These scratches do not run across the sides. On the outer face are several parallel transverse grooves, quite shallow and rounded, in which the scratches are found. These were apparently worn into the bone. There are a few less prominent grooves on the inner face. The grooves are slightly curved at the bottom along their length, bending slightly with the curvature of the tool face. They are sharply terminated, however, and do not curve around onto the shaft sides. These grooves are quite narrow, but smooth and rounded in cross-section. The basal half and base of this tool are very much rounded and polished, as if extensively handled. It is evident that this implement was not twisted in use, but on being inserted in the worked material was moved laterally up-and-down (or side-to-side). Further, this mode of use must have continued for an extended period of time to account for the marked degree of wear, and must have been performed in a very uniform manner as nearly all the wear is found on one face.

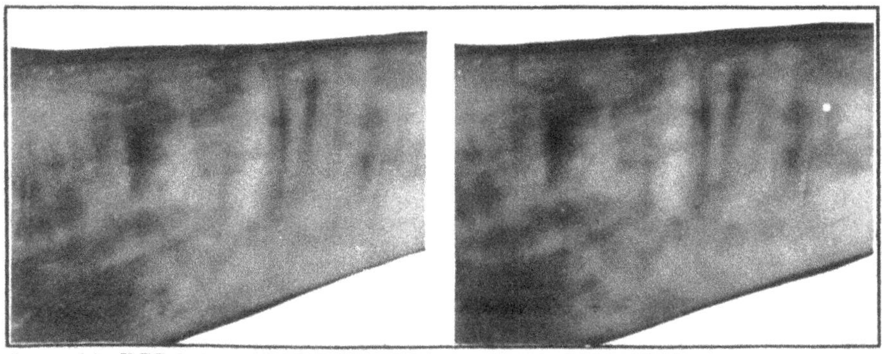

Plate 66: KEB 5: Stereogram of Worn-in Grooves; Face View (10X)

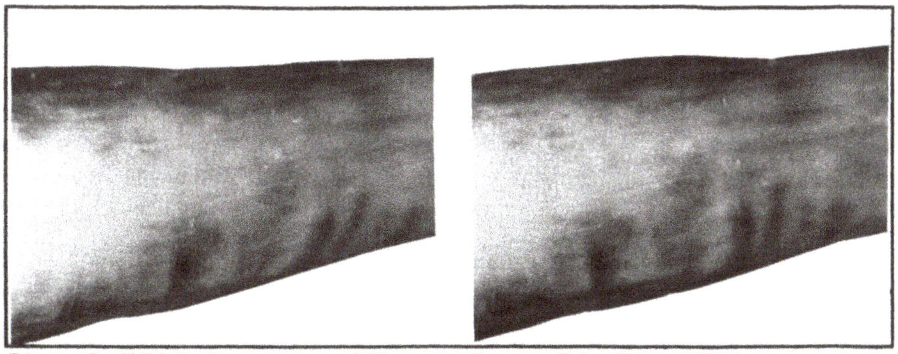

Plate 67: KEB 5: Stereogram of Worn-in Groves; Edge View (10X)

The smooth, uniform curvature of the worn grooves indicates that the worked material was soft and yielding. The narrowness of the grooves suggests that they resulted from friction with fiber or cordage, which would have tended to slip to the bottom of the grooves and resulted in their continual deepening. It is tempting to speculate that this implement was used to press down the weft in the course of weaving, or in performing an analogous task in the manufacture of finely-worked basketry. The observed wear on the tip could have been caused on its insertion between the closely-spaced parallel fibers of the warp, and the friction of the warp may have produced the observed grooves as the tool was moved up-and-down to press down the weft. The asymmetry of wear would have resulted from the continual use of the tool with the same hand.

Fish Vertebral Spines: Two specimens, KEB 32 and KEB 33, consist of the vertebral spines of fish. These are unworked, but show signs of use. The vertebral foramen at the base of KEB 32 forms a natural perforation for the tool, making it resemble a sewing needle. There is no wear visible in this hole, however, and virtually none at the basal end. The small, sharp transverse processes of the vertebra remain intact, and these would surely have been removed (or at least worn down) had the implement been passed completely through the worked material in the manner of a sewing needle. The tip, 0.3 mm in diameter, is slightly chipped and rounded, but light rotation scratches are visible surrounding the tip at a distance of 0.7 mm from the tip end. KEB 33 is similar, but the vertebral foramen is incomplete at the base. The base shows some rounding, perhaps from handling. Rather prominent transverse processes remain on this specimen as well. The tip, 0.4 mm in diameter, is complete, well rounded, and surrounded by a few light rotation scratches at a distance of 3.5 mm from the tip. It is unlikely that these specimens could have served any other purpose than as fine perforators, or perhaps as pins.

Possible lithic retouchers: A total of 12 of the 160 points and implements discussed in this section show scores running either across the tip (two), transversely across one side of the shaft within a few millimeters of the tip (six), or both (four). Five of these scored specimens are parallel sided implements, one is a narrow implement, five are split metapodials, and one is a fragment. The grooves are fairly deep and usually V-shaped, which are nearly but not perfectly parallel, and somewhat irregularly spaced, tending

Plate 68: WAD 2 (1X)

to form clusters. In one case, WAD 2, these scores are associated with a transverse notch about 12 mm from the tip. Five of these specimens also show light rotation scratches, one shows heavy rotation scratches, and two show transverse scratches across the tip.

As these scores are confined to the tip area they are surely not the result of accident, nor are any of them the distinctive marks left by rodent gnawing. They are not confined to any specific form of implement, but all occur on fairly small, narrow specimens, the largest of which is HAY 25, previously discussed. As they occur on only 7.5% of the points and pointed implements they are very likely to represent an occasional atypical rather than normal use of these tools.

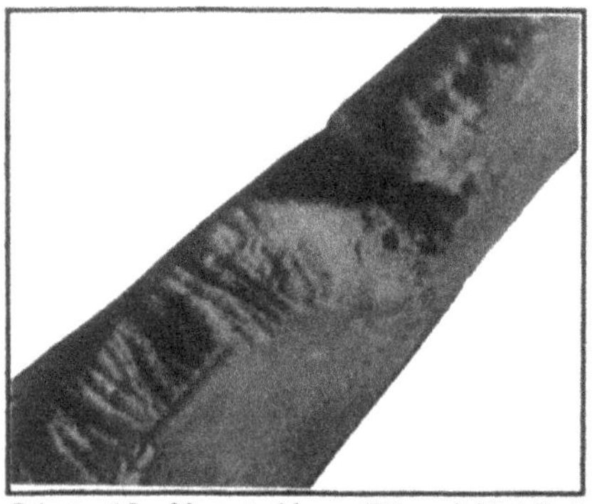

Plate 69: Notch Near Tip of WAD 2 (10x)

These small implements may be compared with a very much larger complete implement, KEB 15 (150.5 mm long, made on a large long bone fragment). This tool has a large, irregular wedge-shaped tip, approximately 6.4 mm across. The implement shows very little rounding or polish. Numerous deep, sub-parallel scores run across the ridge of the wedge-shaped tip, and there are numerous other scores running transversely across the tip to about 18 mm back from the tip.

It is very likely that all these tools were used for retouching the edges of flint implements. KEB 15 may well have been used in the process of pressure-flaking. The small tips may have been used in the notching and retouching of the thin edges of blades and microliths. The wear appearing on these specimens may be compared to the very similar pattern to be found on experimental specimens used to retouch a sharp flint edge (Chapter 5).

Fragments:

Finally, there are twenty-two very small tip fragments that may have been broken from any of the major groups above. With one exception these will receive no further comment, but the wear appearing on these specimens will be included in Table 4. HAY 28 is somewhat unusual in that the tip area of the shaft appears to have been worn by rotation to a regular conical shape, while the sides of the shaft are nearly parallel. At the farthest extent of this worn area there is a clear worn-in step or shoulder; farther up on the shaft the sides are quite irregular in shape. The conical area is surrounded by many fine, parallel rotation scratches. The presence of a worn-in step has previously been noted in the large implement fragment HAY 27. The tip of HAY 28 is 1.8 mm in diameter and moderately rounded, although missing a large chip. The appearance of this step suggests that these tools were used for perforating or enlarging holes in a fairly firm but yielding material, perhaps wood.

Table 4 summarizes the occurrence of scratches and scores for the entire sample of 160 points and pointed implements with small, rounded or sharp tips (KEB 15, KEB 43, and KEB 5, all of which have unique characteristics, are not included).

TABLE 4

NATUFIAN POINTS AND POINTED IMPLEMENTS: OVERALL SCRATCH PATTERNS

Scratches	Number Worn
Total rotation	73 (45.6%)
Light rotation	44 (27.5%)
Heavy rotation	13 (8.1%)
Angled at tip	16 (10.0%)
Randomly oriented	1 (0.6%)
Transverse	6 (3.8%)
Total scores	12 (7.5%)
Scores across tip	2 (0.1%)
Scores across shaft	6 (3.8%)
Scores, both tip and shaft	4 (0.3%)

Double-Pointed Implements

Natufian sites have produced a number of bone implements, of greatly varying sizes, which are similar in that they are all narrow objects pointed on both ends. They are grouped together here purely as a matter of convenience.

Small double-pointed objects

During the excavation of the site of El Wad some small double-pointed objects were found. They were described as follows: "These are tiny slips of bone, pointed at both ends, which were probably used as gorgets in fishing . . . ." (Garrod and Bate 1937: 37).

This functional speculation has become generally accepted in the literature. Superficially similar objects used in fishing are known from various ethnographic contexts, such as the Eskimo and modern Finland (J. G. D. Clark 1952). Garrod's interpretation, however, was based on only three specimens. Other uses for these objects are possible and must be considered.

This section will discuss in some detail a sample of thirty-four similar objects. There is considerable variation in form within this group. In addition there are two specimens, HAY 120 and HAY 124, which may be set apart from the others in that they have one rounded rather than pointed end, and one of them is "decorated" with three raised ridges running around the center of its shaft.

Generally speaking, these objects all have the following characteristics in common. Complete specimens have a length between 20.0 mm and 45.6 mm and a central width between 1.8 mm and 4.3 mm (there is one broken specimen with a central width of 1.1 mm and one with a central width of 4.9 mm). They range in thickness between 1.0 mm and 3.7 mm. The ratio of thickness to width is quite variable, ranging from 0.33 to 0.95 (from quite flat to nearly round). The great majority are about 1.5 times as wide as thick. Although one is tempted on direct observation to regard the flattened and more rounded specimens as variant types, the thickness to width ratio is nearly normally distributed, suggesting that this is not warranted. Based on complete specimens only, the ratio of length to width ranges from 7 to 14.6. If one includes the estimated original lengths of broken pieces the range was probably somewhat greater, from 6 to 16. For complete specimens both thickness and width are strongly correlated with length ($r = 0.749$ and 0.772, respectively), but the central area, estimated by thickness multiplied by width, is even more strongly correlated ($r = 0.806$). That is, for a given length, a thinner than average specimen is also wider than average. The variation in width is therefore likely to be a mechanical effect arising from the need to maintain the shaft at optimal area and strength. Considering the very small size of these objects, it is likely that much of the variation in shape as opposed to size is unintentionally due to inaccuracies of fabrication. Some of this variation must also be the result of inaccuracies of measurement.

Asymmetrical specimens: With two exceptions, all complete specimens are roughly symmetrical about their midpoints, with two pointed tips. The two exceptions, mentioned above, have one rounded end, but as they fit well into the range of dimensions above they are included with this group. One of these, HAY 120, is rounded in cross-section and has one rounded and one pointed end. The other, HAY 124, is similar but smaller, and is surrounded at its midpoint by three raised ridges. It seems clear that these two should be set aside as a sub-group. Whether this might be justified in light of the wear-patterns will be discussed below.

Specimens with midpoint scores: Three specimens (one complete, two broken) have several score lines cut transversely across their surface at midpoint. The complete specimen, HAY 52, has four parallel cuts across the outer surface (the surface which had been the outer surface of the bone). Three of these are close together and fairly shallow, the fourth is slightly

Plate 71: Scores Across Central Shaft of HAY 52: Face and Edge View (14X)

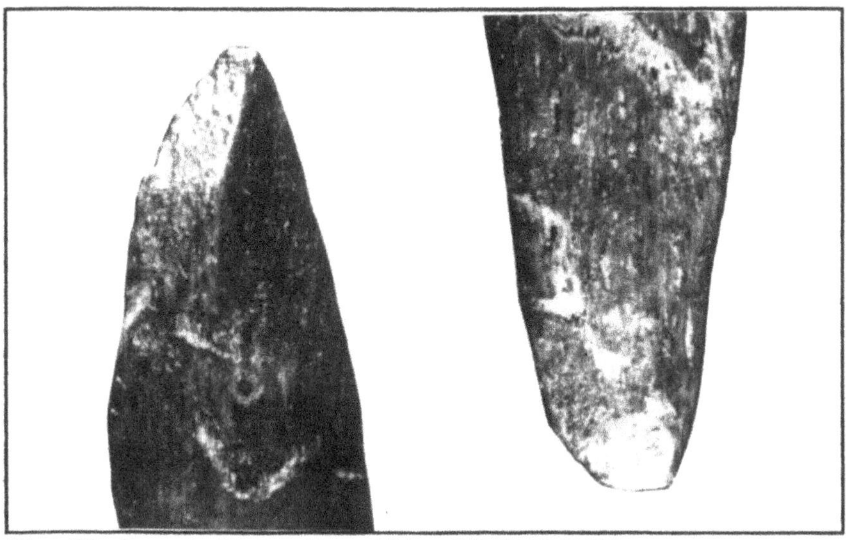

Plate 72: Opposite Wedge-Shaped Tips of HAY 52 (14X)

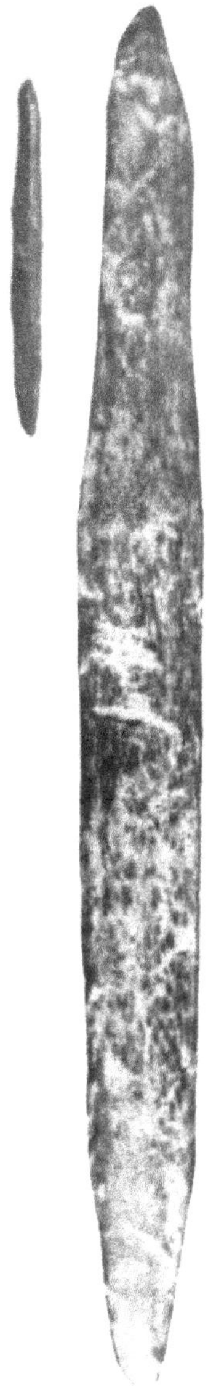

Plate 70: HAY 52: (1X & 4X)

deeper. They have V-shaped bottoms, suggesting that they may have been made with the cutting edge of a flake. They run across the face at an angle of about 75° to the long axis of the specimen. One of the broken specimens, HAY 1, also has three cuts running at nearly right angles across the outer face only; it was probably the midpoint of the complete specimen. These are not as clearly marked as on the complete specimen and are somewhat obscured by calcitic deposit. The third specimen, HAY 196, is rather deeply scored across both the inner and outer faces. There are three deep cuts with V-shaped bottoms running parallel at angle of about 75° to the long axis on both faces. This specimen, and perhaps HAY 1 as well, appears to have been broken along a fourth score mark. These three specimens are rather similarly shaped, showing largely unworked faces with most of the shaping confined to the tips. Their dimensions are rather similar, with estimated length ranging from 35 to 38 mm, width from 2.7 to 3.4 mm, and thickness from 1.5 to 2.1 mm. HAY 196 has been charred black, HAY 52 is darkened, and HAY 1 is the natural color of the bone.

Plate 73: HAY 1: (4X & 1X)

Plate 74: HAY 1: Wedge-Shaped Tip & Scores Across Central Shaft (14X)

Plate 76: Scores on Opposite Faces of HAY 196 (14X)

Plate 75: HAY 196 (4X & 1X)

It is of considerable interest that two of these specimens were found in fairly close proximity in Hayonim Cave. HAY 52 and HAY 1 were both found above Burial IV; HAY 196 was about four meters away from these specimens.

Other specimens: The remainder of the specimens show no special variations in form.

Tip size: The tips of these implements are all moderately small, ranging from 0.15 mm to 1.3 mm in diameter (one possible small double-point has a tip diameter of 1.8 mm). On the 31 assured double-points there are 43 complete tips (including 12 pairs), with a mean tip diameter of 0.69 mm, s = 0.34. The two pin-shaped objects have tip diameters of 0.7 mm and 0.8 mm; the possible double-points have tip diameters of 0.35 mm, 0.45 mm, and 1.8

mm. This last might possibly be the rounded end of a pin-shaped specimen. On specimens where both tips remain they are often but not always nearly the same size.

Tip shape: The majority of the 53 tips remaining are hemispherical in shape, four are somewhat ovoid, and 11 are irregular. The pin-shaped objects both have round tips, as do two of the possible small double-points; the third is irregular. Of the specimens with incisions across the midpoint, HAY 1 has an ovoid tip, HAY 196 has a wedge-shaped tip, and on HAY 52 both tips are wedge-shaped.

TABLE 5

SCRATCH PATTERNS ON NATUFIAN SMALL DOUBLE-POINTED OBJECTS

TIP-BY-TIP

Scratches	Number worn	Proportion
Light rotation	5	9.4%
Heavy rotation	4	7.4%
Transverse	3	5.7%
Angled at tip	2	3.8%
Total rotation	11	20.8%

NOTE : The total number of tips is fifty-three.

SPECIMEN-BY-SPECIMEN (N = 31)

Scratches	Number worn	Proportion
Light rotation	5*	12.9%
Heavy rotation	3*	9.7%
Transverse	3	9.7%
Angled at tip	2	6.5%
Total rotation	9**	29.0%

*One specimen has scratches on both tips.
**Two specimens have scratches on both tips.

Wear-patterns: Microscopic examination reveals that many of these objects show recognizable wear-traces. Most of the specimens show some degree of rounding and/or polish on their tips. A substantial proportion show scratches surrounding their tips which suggest that the implement was rotated or twisted during use.

The proportion of tips indicating tool rotation is considerably smaller than that found on most pointed implements but is far too large to be the result of accident or atypical use. On a specimen-by-specimen basis the proportion of 29% which have been twisted in the course of use is clearly significant. Two of the specimens with incisions at the midpoint have transverse scratches running across the faces of the wedge shaped tip (HAY 52, HAY 196). None of the possible small double-points show scratches. One of the pin-shaped objects, HAY 20, has heavy rotation scratches round the tip.

Plate 77: HAY 130 (1X & 4X)

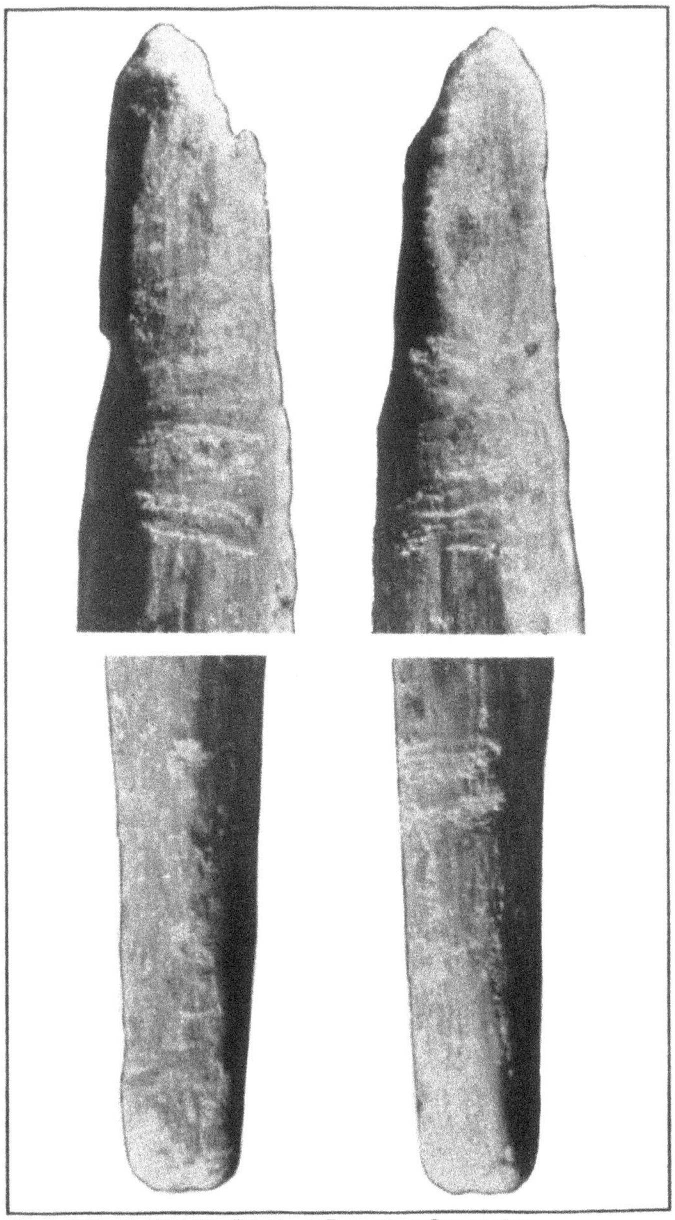

Plate 78: HAY 130: Distinct Rotation Scratches Surrounding Both Tips (14X)

Plate 79: Distinct Rotation Scratches Surrounding Tip of HAY 122 (14X)

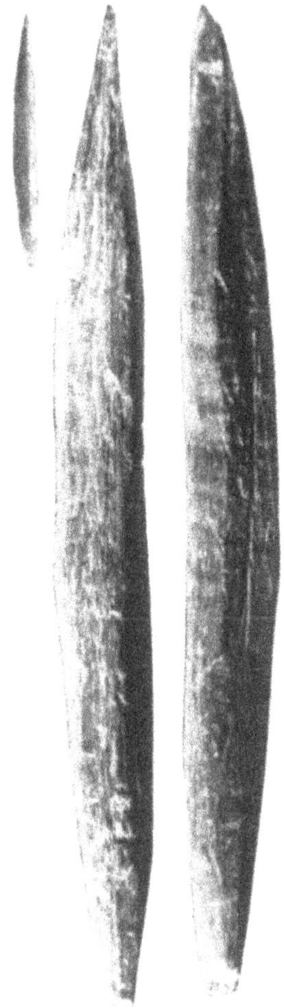

Plate 80: HAY 122 (1x & 4X)

Conclusions: It is clear that the interpretation of these implements as gorges for fishing could not be correct, at least for most specimens, as it would be difficult to reconcile this use with the evidence for tool rotation. It is of interest that very light scratches indicative of rotation are to be found on all three of the El Wad specimens. However, the three Hayonim specimens with incisions at the midpoint could possibly have served this as fish gorges. Considering their size, the central incisions are almost certainly functional rather than decorative. Most probably they would have served to help secure a line tied around the implement's central shaft, although no clear wear from such a line could be found. The wedge-shaped tips, showing slight rounding and transverse scratches running across the faces, would not necessarily contradict the use of these tools as fishing gorges. A more mundane interpretation would also fit the evidence however; they may have served as buttons or fasteners.

The simplest interpretation for the remaining specimens would be that they were used to pierce a soft material in the manner of the points and pointed implements of the preceding section. If this is so they may have served as perforators, but in view of their size they are more likely to have served as pins or fasteners. One may not easily make this distinction on the basis of tip wear, just as it would be quite difficult to distinguish between a metal needle and a pin on this basis. It may be noted however that the central shaft of these implements is not very heavily rounded and polished. Extensive rounding would be expected had they been extensively handled as a needle would have been.

The elevated proportion of specimens with heavy rotation scratches is surprising (in comparison with the simple points of the preceding section). This may be caused by tool rotation against a stiff material such as wood; this leads to another possible interpretation of function. It is possible that these objects may have been twisted into wooden hafts, possibly serving as barbs (see section on barbed points). Three circumstances argue against such an interpretation: (1) if these objects were barbs a much higher proportion of heavily chipped and broken tips would be expected than is observed, and the wear on the two ends would be expected to be dissimilar, which it generally is not; (2) one specimen shows such heavy scratches at both ends; and (3) one of the pin-

Plate 81: HAY 120 (1X & 4X)

Plate 82: Pin-Shaped Object HAY 120; Distinct Rotation Scratches Near Tip and On Central Shaft (14X)

shaped objects, which would not have served as a barb, shows similar scratches.

Large, narrow double-pointed implements

Four complete and three broken specimens may be included in this group. It is not actually a homogeneous category as the specimens vary widely in size. Therefore it seems best to describe the specimens individually. The four complete specimens range from 53.0 mm to 177.0 mm in length, and from 3.6 mm to 7.9 mm in central width.

The smallest specimen, SHK 9, (53.0 mm long, 3.6 mm wide, 3.2 mm thick) is only slightly larger than the largest of the specimens included in the group of small double-points, and its segregation from this group is somewhat arbitrary. It has a complete rounded tip on either end, both of which show moderate use-rounding but no chips. The tip diameters are 0.45 mm and 2.0 mm and light rotation scratches were observed about 3.2 mm from the smaller tip. There is a moderate polish over the surface. This specimen may well be similar to the pin-shaped objects described above and may have functioned similarly.

KEB 1 (74.6 mm long, 4.1 mm wide, 3.0 mm thick) is the next smallest specimen. It is similar to SHK 9 in that it has two round, somewhat worn but unchipped tips, 0.6 mm and 1.4 mm in diameter. As with SHK 9, the smaller tip is encircled by light rotation scratches which are visible to 20 mm from the tip. The shaft is polished. Apart from the larger size this specimen is very similar to SHK 9.

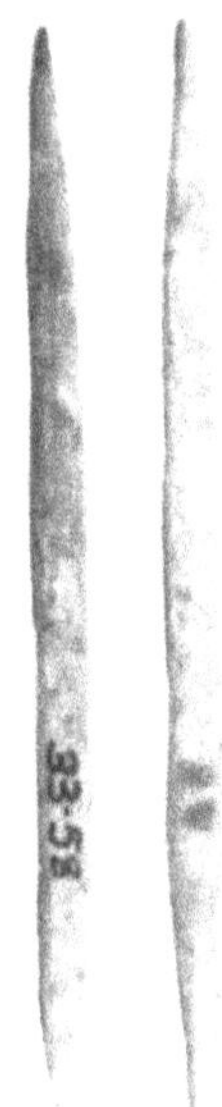

Plate 83: Large Double-Pointed Object KEB 1 (1X)

ORN 4 is very much larger (150.2 mm long, 7.7 mm wide, 3.8 mm thick). Both tips are complete and unchipped but dissimilar in size (0.6 mm and 4.5 mm in diameter), so that it is similar in shape to the pin-shaped objects but larger. It shows signs of handling and rounding and polish at the tips but no scratches.

KEB 6, the largest specimen (177.0 mm long, 7.9 mm wide, 7.9 mm thick) differs from the others in that the shaft has been quite accurately rounded. The central shaft area is covered with many patches of fine parallel scratches which would appear to the result of intentional smoothing or polishing. The implement has one complete and one missing tip. The end of the shaft with the complete tip (0.6 mm in diameter) is markedly more polished than the opposite end. The tip is encircled by fine rotation scratches to a distance of 30 mm from the tip (the shaft is 6.5 mm wide at this point).

In view of the great similarity of wear-patterns seen on these implements it would seem that they all served the same basic purpose, although in different contexts (rather analogous to the different uses of a set of screwdrivers or chisels of different size). They are all clearly primarily perforating implements.

Another broken specimen, KEB 45, appears to have originally been about 120 mm in length. The remaining tip is ovoid in shape and quite large (5.0 mm in diameter). A flattened area is visible which may be worn in but is more probably an accident of manufacture. Randomly-oriented scratches are visible at this tip. The most likely interpretation of this implement is that it is the base or handle end of a single-pointed tool similar to those above, although this is of course speculative as the other tip end is missing.

Broad double-pointed implements

HAY 147 and WAD 18 are implements of bone with roughly similar dimensions. HAY 147 is a nearly complete unworked fragment of a long bone shaft with two crude tips. WAD 18 is unusual for a Natufian implement in that it has been worked to shape by grinding. It has one rounded and one rather spatulate point, 1.5 mm and 2.7 mm wide, respectively.

Implements of this form, specifically those made from antler, have occurred in Near Eastern contexts from quite early times, being found in small numbers in Upper Paleolithic and Kebaran assemblages. They are usually referred to as "bipoints", and they are generally interpreted as projectile points. The wear to be seen on these implements tends to support this interpretation.

Spatulate Implements

As described above, a number of different forms are included in this category, united only in that these tools all possess a broad, rather than pointed, working end. An attempt has been made to measure these working surfaces, but as their shape is very variable this is intended only as an approximate comparative guide. Two typical (generalized) spatulate tips are illustrated in Figure 6(1). Viewed perpendicularly to the widest dimension, some tips have an approximately elliptical profile, while others have widened blades, shaped rather like screwdrivers. To reflect this difference in shape two measurements have been taken: (1) the tip width, which is taken from the point at which the sides are perceived to turn in to form the tip; (2) the blade width, which is the approximate width of the straight portion of the blade.

Large, elliptical implements

Probable hide rubbers: These are robust implements with elliptical profiles. Three Natufian specimens, all from Hayonim Cave (HAY 47, HAY 117, HAY 119), fall into this group. They are all of antler or horn core. HAY 47 and HAY 119 are represented only by the tip portion; they are quite large (tip width = 13.2 mm and 14.5 mm, respectively). HAY 117 is smaller (tip width = 5.2 mm). It is a complete implement on a horn core, 75.7 mm long. The interior of the base has been hollowed and several transverse grooves cut around the base. The implement may have been hafted, perhaps by inserting a wooden haft into the hollow and wrapping a binding cord around the haft, it being secured to the implement in the cut grooves.

The tips of all these implements are slightly to moderately polished and rounded by use. There is little chipping of the edge of the tip. All three specimens show clear but very fine, rather evenly spaced parallel scratches running back axially from the tip edge, superimposed on the polish. These appear on one face of HAY 47 and HAY 119 and on both faces of HAY 117. HAY 119 shows traces of red pigment, probably ocher, impressed into the surface near the tip.

Very large antler implements with a flattened face

Probable hide rubber / compressors: Four Natufian specimens of antler, HAY 50, HAY 191, WAD 14, and WAD 15 fall into this category. An additional specimen made from a portion of the shaft of a very large long bone, ORN 8, is rather similar is shape. These specimens, particularly those of antler, have very large, broad, smoothed surfaces on one face but are very largely unworked otherwise (except for cutting off the antler section).

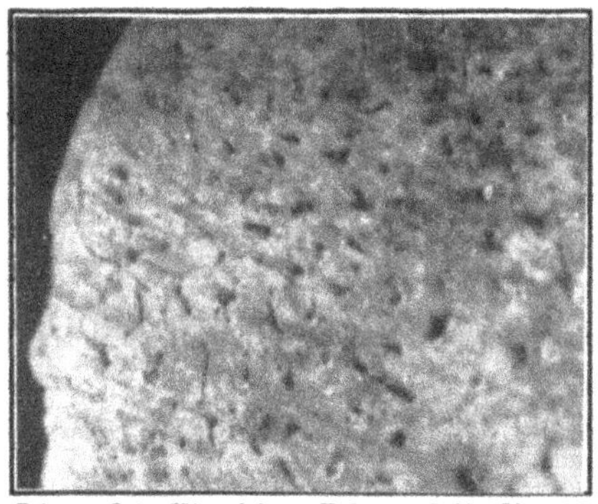

Plate 84: Working Edge of HAY 119 (10X)

Plate 85: HAY 119 (½X)

HAY 50 has the largest smoothed surface, about 25 mm wide by 90 mm long. This surface is slightly polished, and many fine, evenly spaced parallel scratches may be seen with the microscope, running the length of the tool. The face opposite this surface, near the tip, is also polished, and a patch of fine parallel scratches is visible, running at an angle of 45° to the tool axis.

HAY 191 is similar, but is made from a longer, narrower section of antler, and the smooth surface is smaller, about 18 mm by 30 mm. The very tip portion has been broken off. This implement also shows polish and fine parallel axial scratches on the smoothed surface.

WAD 14 is a very large implement made on the pediment portion of a red deer antler (182 mm long, 35 mm wide, 32.2 mm thick at the base). The smoothed surface in this case is about 25 mm by 80 mm. This specimen has little polish. The surface is somewhat roughened, and fine scratches could not be observed. WAD 15 is similar to WAD 14 but even larger (193.6 mm long, 45.5 mm wide, 40.8 mm thick at the base). The smoothed surface is about 32 mm wide by 105 mm long. The microscope shows numerous axial scratches on this surface, but confined to an area 10 mm from the tip. These are not strictly parallel.

The bone specimen, ORN 8, shows no clear scratches, although there is very slight use-rounding and polish of the tip edge.

Plate 86: Probable Rubber / Compressor WAD 15 (½X)

Of the eight Natufian implements in the two groups described above, six specimens (75%) show scratch patterns indicating that they were pushed back-and-forth parallel to their axes, the smoothed surface being pressed against a material which was sufficiently abrasive to leave very fine scratches, but not so coarse as to form deep scratches. As the surfaces of these tools were smoothly curved, without gouges, chips, or particularly, worn-in flats, it is evident that the material was quite yielding. On five of the specimens the long, straight scratches indicate that the tool was moved with rather long, sweeping movements; on the sixth, WAD 15, the wear seems largely confined to the tip and this tool was not moved so regularly back-and-forth as were the others. This is not surprising considering the very large size of this implement.

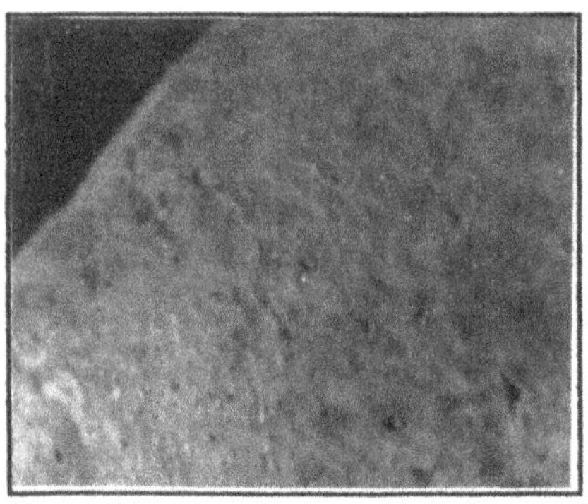

Plate 87: Fine Scratches on Face of WAD 15 (10X)

These implements are usually interpreted as rubbers or dressers for hides, and the wear observed strongly supports this, as does the appearance of pigment which may have been rubbed into the hide surface. The use of silt, clay, or ocher for the smoothing or dressing of hides could easily have resulted in the fine scratches which are observed. Experiment has shown that even a clean leather surface, if it is unlubricated, can result in similar scratches on a bone surface if the bone is rubbed against it with sufficient pressure for an extended period of time.

Another possible use for these tools, as picks, wedges, or mattocks, can be ruled out, as the tip edges show very little of the chipping, battering, or deep scratches as would be expected from such use, and the observed scratches could not be explained in this way.

Bone implement with small spatulate tip

There is one Natufian specimen, HAY 26, made from bone, which has a small spatulate tip. To some degree this specimen resembles similar Zagros

Protoneolithic artifacts described in Chapter 7. The tip, with a blade 1.9 mm across, is heavily chipped on one side and bears scratches across it. This does not seem to have served as a hide dresser, although its actual function is indeterminate.

Bodkin-like implements

Probable netting or matting tools: SHK 1 is a bodkin-shaped implement made on a rib. It is a complete tool (71.2 mm long, 15.5 mm wide, 2.8 mm/ thick) with a flat tip with an elliptical profile and a perforation at the base (hole diameter = 3.7 mm). This tool could hardly be interpreted as other than a bodkin for use in netting or mat-making. The tip is unchipped and well use rounded and polished, with the heaviest polish running up the sides of the shaft. The hole is heavily rounded and polished on its edges. There is no marked asymmetry in its wear.

Plate 88: SHK 1 (½X)

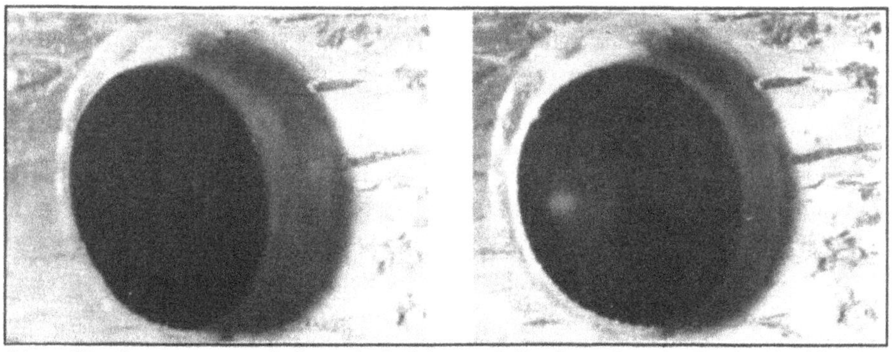

Plate 89: Stereogram of Rounded Edges of Perforation in Probable Matting Tool SHK 1 (10X)

WAD 12 is also a flattened implement (71.2 mm long, 10.9 mm wide, 5.2 mm thick) with a perforation at the base (hole diameter = 2.9 mm). The tip is irregular in shape but roughly rounded and small (1.2 mm in diameter). The very tip appears slightly battered but it is otherwise well rounded and polished. The sides of this tool are also polished as in the above specimen. Scratches cross the shaft faces transversely but are not prominent near the tip. There is no wear visible around the edges of the hole.

KEB 11 is a very long, flat implement, made from a rib and consequently curved. (192.0 mm long, 10.0 mm wide, 3.8 mm thick). It has a rounded tip. There is a perforation, 4.4 mm diameter, at the base. There are no chips on the tip, which is well rounded and polished. There is some slight use-rounding and polish on the sides and on the outer convex face. Very few scratches are visible on this specimen. There is almost no wear visible on the perforation edges.

Despite the difference in length WAD 12 and KEB 11 are otherwise quite similar, both is shape and wear-pattern. The wear at the tips indicates that the implement penetrated a worked material providing some resistance,

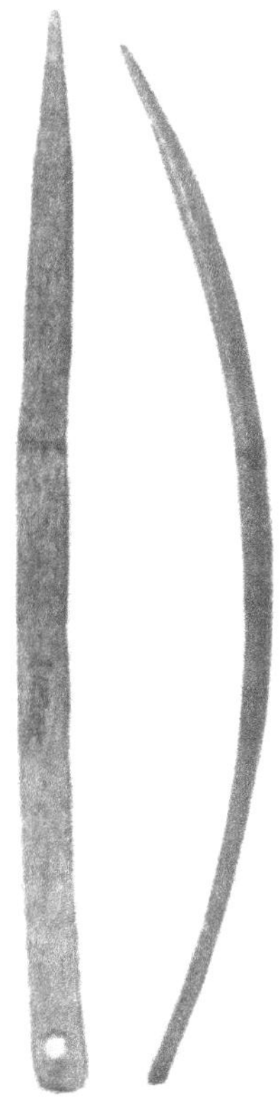

Plate 90: KEB 11 (½X)

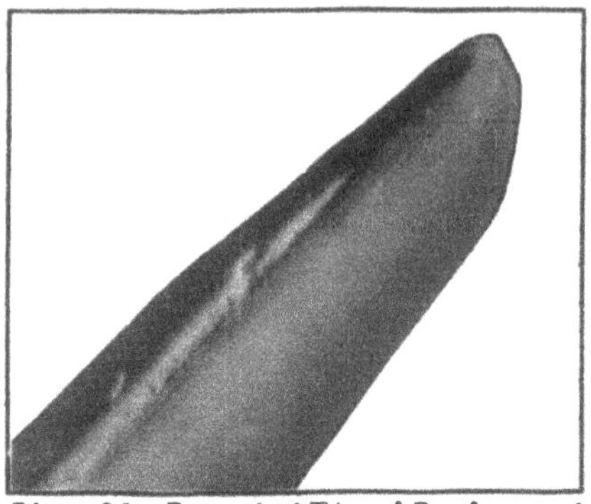

Plate 91: Rounded Tip of Perforated Rib KEB 11 (10X)

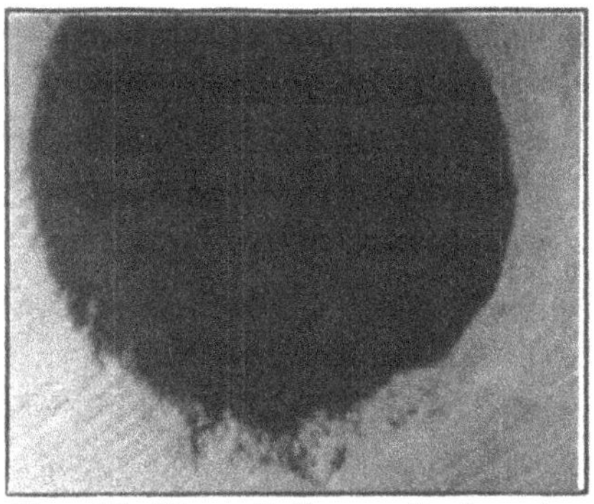

Plate 92: Virtually Unworn Perforation at Base of KEB 11 (10X)

perhaps closely spaced matting or weaving. The lack of wear on the edges of the perforations is surprising, considering the heavy wear seen in most other specimens (see Chapter 7). It could perhaps be explained if the cord drawn by the tool was fastened at his point rather than drawn loosely through the hole.

Implement with perforation at tip (lacing tool?): Perhaps the most interesting specimen in this group is SHK 4. This is a flat implement made on a rib (56.9 mm long, 12.2 mm wide, 3.6 mm thick). It has a small, round, rather sharp tip, 0.6 mm in diameter. This tool has a perforation (4.5 mm in diameter) placed very near the tip. The center of the hole is 13 mm from the tip. The tip is unchipped and is completely rounded by use and polished. The tip is surrounded by fine rotation scratches which are visible along the sides of the shaft from the tip to 13.6 mm form the tip, that is, even with the hole. The interior edges of the hole are rounded by use and polished and are slightly more worn on the side of the hole nearer the tip.

From the small worn tip and the rotation scratches it is evident that this tool was forced, with a twisting motion, through compact material such as hide, or a tightly woven material or matting. A lace may have been pushed through the hole. Alternatively the implement may have been pushed through the worked material and the lace then threaded though the hole, after which the tool was withdrawn. This would explain the asymmetry of tool

Plate93: SHK 4 (½X)

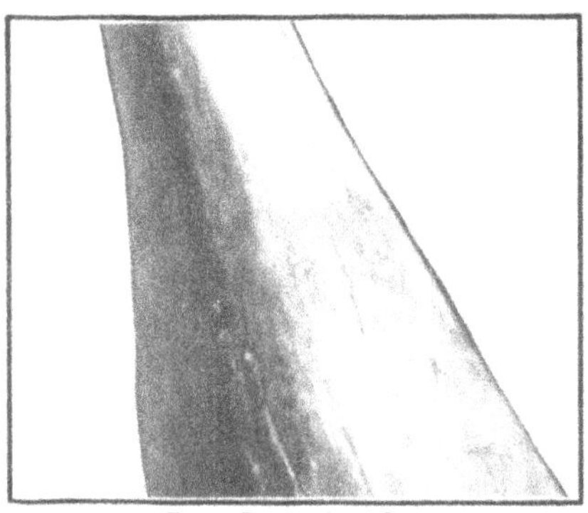

Plate 94: Fine Rotation Scratches (At Limit of Visibility in Reproduction) Surrounding Tip of SHK 4 (10X)

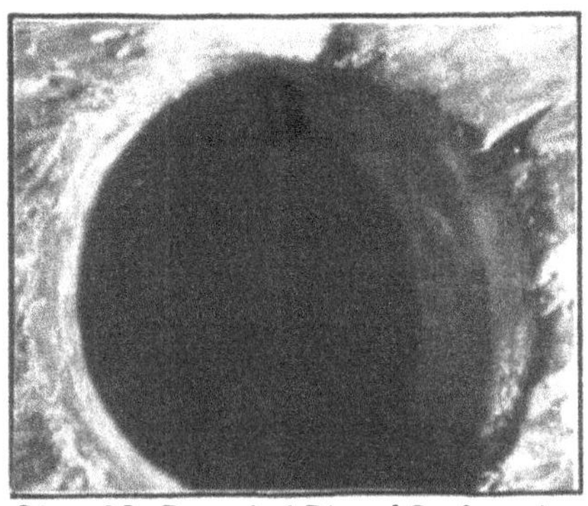

Plate 95: Rounded Rim of Perforation Near Tip of SHK 4 (10X)

wear. A third possibility might be that a loop of lace was pushed through the worked material and caught by a second lace from beneath.

Other implements: KEB 26 and KEB 27 are similar in shape. They are complete flattened implements with broad tips and no perforation. KEB 27 (74.1 mm long, 13,1 mm wide, 2.6 mm thick) is made on a rib. KEB 27 (77.2 mm long, 19.4 mm wide, 6.3 mm thick) is made on a long bone shaft fragment. Neither of these implements show much wear, with no chips at the tip and only slight rounding and polish. There are a few transverse scratches across the tip of KEB 26. There is too little evidence to make any judgement as to the use of these tools.

Barbed Points

Barbed points, usually called "harpoons" occur at several Natufian sites. There is no reason, however, to assume that these points have been used in fishing as the name harpoon might suggest. They may have been used in any hunting activity. It is scarcely likely that these distinctive implements could have served any other purpose. It is worthwhile, however, to examine the wear on these specimens, both to confirm the hypothesized use and to provide a comparison with other implements.

Eighteen specimens are included in this sample. Of these six are complete; that is, both tip and base are present. All the specimens have a single row of barbs. They range in length from 42.7 mm to 150.6 mm. KEB 34 is considerably larger than the others and appears to be made of antler while the rest are of bone. It will be discussed separately below.

Small barbed points: The number of barbs on the other complete specimens is variable, ranging from one to four. Among the broken specimens three are missing the tip (one to two barbs remaining), four are missing the base (one to four barbs remaining) and five are represented by shaft fragments (on to four barbs remaining). Besides KEB 34 there are six unbroken tips. In general the very tips of these barbed points are not round but slightly flattened, being rather ovate in cross-section (tip width 0.5 mm to 2.0 mm). Most have a battered appearance. The wear appearing on these tips is summarized in Table 6. None of these tips show any distinctive scratches. The high proportion of heavily chipped tips (57%) is as would be expected from the impact of projectile points.

TABLE 6

TIP WEAR ON NATUFIAN BARBED POINTS

	Number of Tips
Rounding	
None	2
Slight	4
Moderate	1
Polish	
None	2
Slight at tip	4
Tip well polished	1
Chipping	
None	1
Slight one side	1
Slight both sides	1
Heavy one side	3
Heavy both sides	1

Most of the other wear on these implements is seen on the tips of the barbs. Some of these have been broken off completely. Most of the complete barbs show some degree of rounding due to use, particularly on the outer side of the backward-facing barb. The outer edge of the barb is often quite markedly polished. Some barbs show fine scratches, which usually run transversely across the edge of the barb. For the barbs overall, see Table 7. The wear on these barbs tends to be more marked the farther the barbs are from the tip. This effect may be related to the widening of the shaft

Plate 96: KEB 35 (IX)

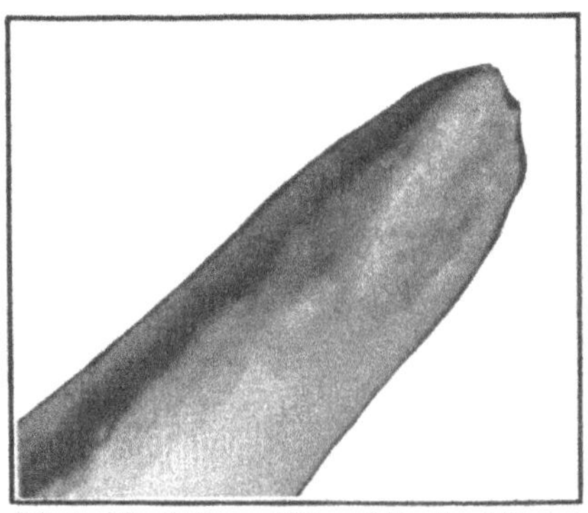

Plate 97: Chipped Tip of Barbed Point KEB 35 (10X)

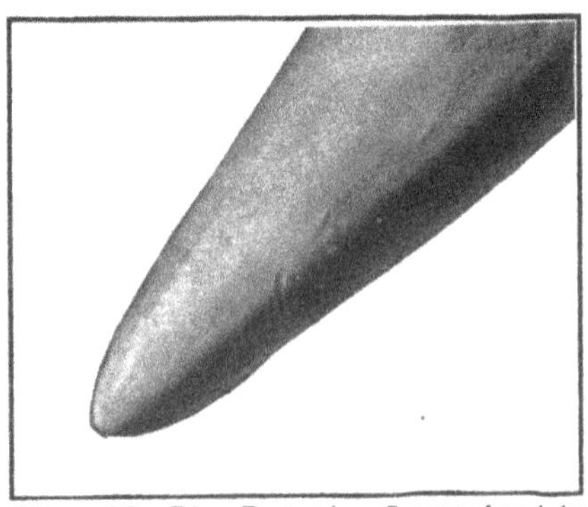

Plate 98: FineRotationScratches(At Limit of Reproduction) Around Base of KEB 35 (10X)

farther from the tip, leading to greater friction on the barbs farther back from the tip. The wear is probably the result of the penetration (and later withdrawal) of the shaft into its target or into the soil. Tyzzer (1936) has experimentally demonstrated that appreciable polish can be put on a bone point by shooting it into an earthen bank with a bow.

TABLE 7

NATUFIAN BARBED POINTS: BARB WEAR

	Number of barbs
Rounding	
None	9
Slight	15
Moderate	9
Polish	
None	10
Slight at tip	12
Tip well polished	8
Polish up shaft	3
Chipping	
None	26
Slight one side	3
Slight both sides	2
Heavy one side	2
Missing tip	10
Scratches	
None	26
Transverse across tip	9*

*Of which two are seen across broken tips.

In addition to KEB 34 there are seven specimens with complete bases, plus one with only the very end of the base missing. The bases of these implements actually form fairly sharp tips, which may be round, ovate, or rather chisel-shaped, but which are all quite small, ranging from 0.9 mm to 1.8 mm. Most show considerable use-rounding and polish as summarized in Table 8 for complete bases.

Most of the bases show distinct scratches near the basal tip. Of the eight specimens, four show fine transverse scratches situated from 0.9 mm to 5.0 mm from the basal end. Two specimens show clear light rotation scratches at 5.0 mm and 15.0 mm from the base.

TABLE 8

NATUFIAN BARBED POINTS: BASE WEAR

	Number
Rounding	
None	2
Moderate	4
Complete	1
Polish	
none	2
Slight at tip	1
Tip well polished	2
Polish up shaft	2
Chipping	
None	6
Slight one side	1

The most likely explanation of the wear-patterns seen on the bases of these implements would be that they were hafted by forcing the bases, sometimes with a twisting motion, into a haft, probably of wood. If these implements were simply barbed points and permanently hafted this wear would be difficult to account for. It is more likely that these were detachable points, perhaps used in the manner of a true harpoon. Scratches, could of course appear after a single hafting but the rounding and polish observed would require prolonged friction. This implies repeated re-haftings. There is no evidence, however, that these implements were tied to a line as a true harpoon would be.

Possible fishing implement: KEB 34, of antler, is the largest of the barbed points, 150.6 mm long. It has five barbs. The ovoid tip is small (0.7 mm), slightly chipped, and slightly rounded and polished. Unlike the other specimens, the base of KEB 34 is large and irregular. It shows no signs of wear. This implement must have been permanently hafted, perhaps by lashing. The barbs all show considerable chipping, rounding, and polish on their tips, and this wear is most prominent on the barbs farthest from the tip. The second through fifth barbs all show distinct parallel transverse scratches crossing the outer edge of the barb. Unfortunately, no ready explanation can be offered for these transverse scratches. They might perhaps be the result of a casual abrasion. It is rather surprising that the barbs near the base should be so much more heavily rounded and polished than those near the tip; the tip itself is only very slightly rounded, and the first barb is virtually unrounded. If this implement had penetrated a target in the manner of a projectile point one would expect much greater wear of the tip in comparison with the other barbs than is actually observed.

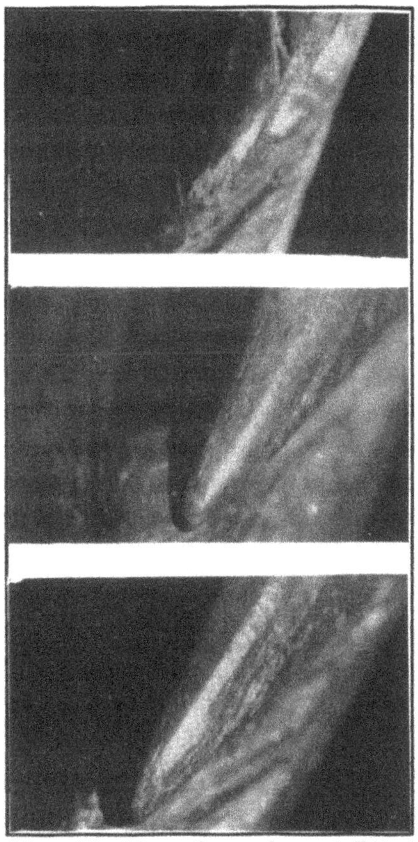

Plate 99: First, Second, and Third Barbs on KEB 34, Showing Increasing Rounding and Polish Toward Base End, and Transverse Scratches Across Second and Third Barbs (10X)

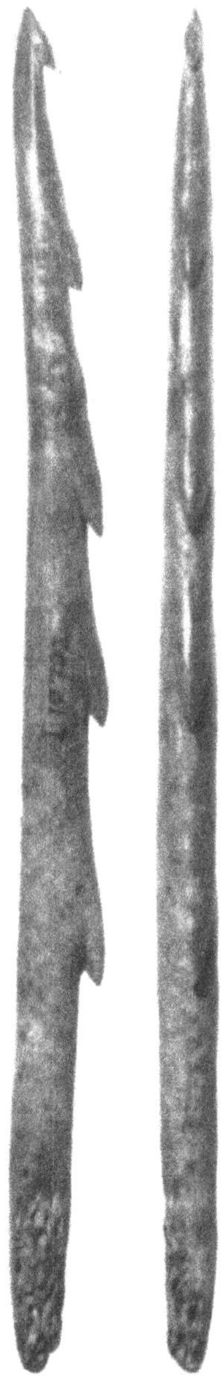

Plate 100: KEB 34 (1X)

It should be observed that the implement is slightly curved, the barbed side being slightly convex, although this might be due to warping on drying. A possible explanation of the observed wear might be that it functioned not as a projectile point but as one prong of a leister, a fishing implement. This is speculative, but leister prongs (found in pairs) which are rather similar are known from European Mesolithic sites which are roughly contemporary (Clark 1952).

Hooks

Kebarah Cave yielded four specimens which have been interpreted as "fish hooks" (Turville-Petre 1932). One specimen, KEB 24, is complete; two, KEB 22 and KEB 25, are complete except that the shaft end opposite the pointed tip is missing. KEB 23 is the shaft portion only, missing the tip. These have all been made by boring a whole in a flat piece and then cutting away one side to form a "U". A tip is sharpened on one leg of the U and a constriction cut around the other leg, evidently to secure a cord. Three specimens are of bone; the fourth, KEB 22, is made on a flat piece of tusk. The wear on these specimens is best described individually.

KEB 22 (33.4 mm long, 21.3 mm wide, 4.9 mm thick) is moderately polished on the side opposite the enamel. The small (0.6 mm) pointed tip is

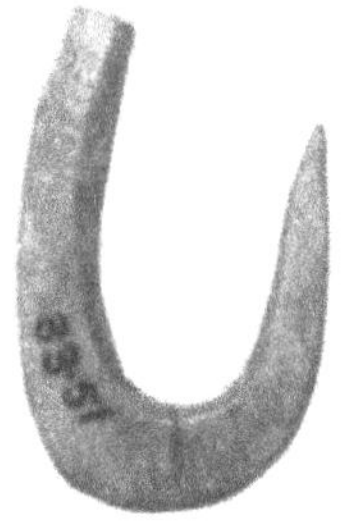

Plate 101: Hook; KEB 22 (1X)

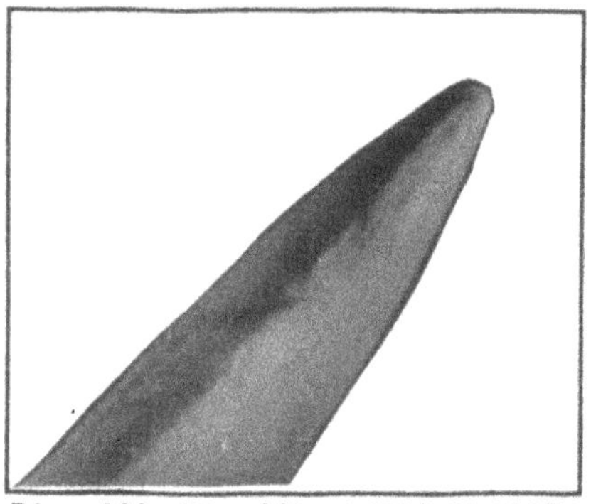

Plate 102: Tip of Hook KEB 22 (10X)

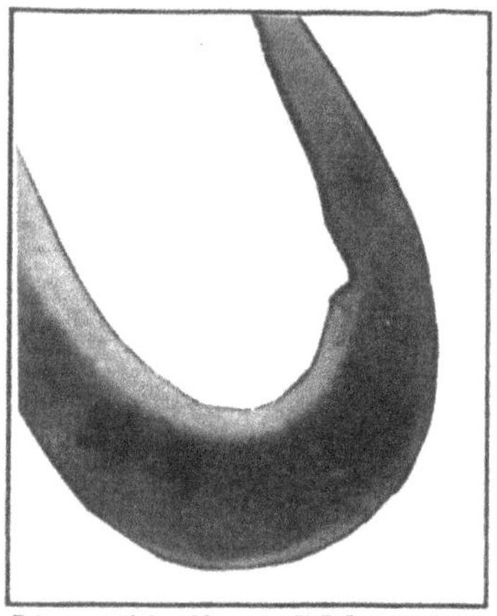

Plate 103: Hook: KEB 22 (2X)

complete and unchipped. It shows moderate use rounding and is slightly polished. The most marked wear near the tip is on the outside edge of the tip leg of the U. Considerable rounding and polish may be seen on the edge of the perforation at the base of the U, particularly on the bony face but also on the edges of the enameled face. Little rounding or polish is to be seen on the shaft leg.

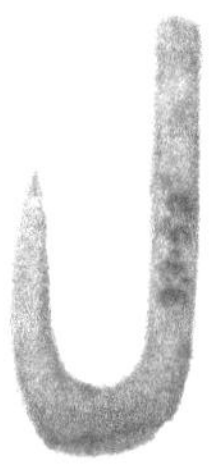

Plate 104: Hook; KEB 24 (1X)

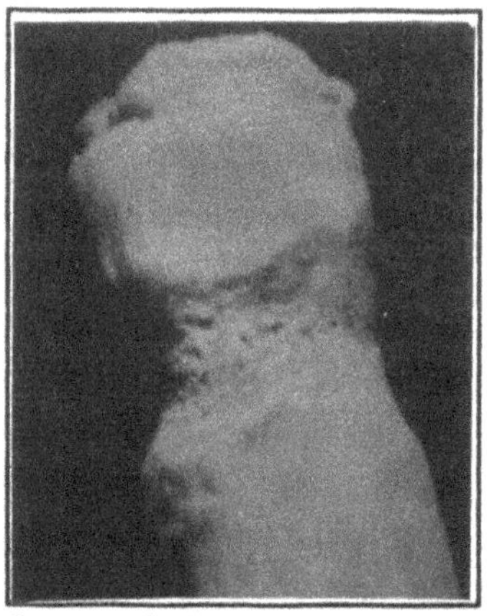

Plate 105: "Matte" Appearing Groove Around Shaft of Hook KEB 24 (10X)

KEB 24 (28.9 mm long, 12.1 mm wide, 2.6 mm thick) is slightly polished overall. The tip (0.35 mm in diameter) is complete, unchipped, and very slightly rounded and polished. The marks left from manufacture are generally very clear on this specimen, and particularly so on the sides of the drilled hole. The edges of the hole, however, at the base of the U show some rounding and polish. The shaft leg of the U shows very little wear but there is slight rounding and polish on the snapped-off end of this leg. The groove which completely surrounds the shaft has a very unusual "matte" appearance which obscures any wear or marks of manufacture. It is possible that this was caused by tearing of the bone surface if a line had been attached and cemented by an adhesive that pulled away. The edges of this groove are slightly rounded and polished.

KEB 25 (24.1 mm long, 13.0 mm wide, 4.1 mm thick) shows very little polish. The small tip (0.3 mm) is unchipped and shows virtually no wear. The marks of manufacture are generally very clear on all of this specimen, including the base of the U and the inside of the perforation. The edges of the perforation and the outer edge at the base of the U show some rounding and polish however. KEB 23 (25.2 mm long, 3.3 mm wide, 2.5 mm thick), a shaft portion only, is lightly polished overall. Manufacture marks are fairly clear. The cut-in notch around the end does not show any rounding or polish to suggest friction from a cord. The very base, however, is slightly rounded and polished. On the inner edges of the shaft, near the point at which it is curved in to form the base of the U which is now missing, there are a number of short transverse scratches.

It is quite difficult to interpret this wear, particularly since the sample is so small. Some observations may be made however. (1) None of the tips, which are all small, show much wear or breakage; (2) the most marked wear appears at the *base* of the U, as rounding of the edges and in one case transverse scratches; (3) the two complete shaft ends both show rounding at the very end, perhaps from handling. It is surprising that the tips are unbroken and show so little wear given the degree of rounding at the base of the U. It seems possible, therefore, that these were not actually fish hooks but fasteners of some sort. If a cord, thong, or belt were caught at the base of the U it might have resulted in the observed wear in this area without appreciably wearing the tip[15]. It must be emphasized that this is merely a speculation. The wear on none of these specimens is sufficiently unambiguous for a clear decision to be made.

Bone Hafts

Numerous bone implements usually interpreted as "sickle hafts" have been found at Natufian sites. Most of the haft specimens are fragmentary; only two nearly complete Natufian hafts are known. One of these, from Kebarah Cave and designated here as KEB 40, is perhaps the most widely known Natufian artifact. It is a large haft, 308 mm long, carved with the head of an artiodactyl on the handle end. A deep V-shaped groove extends up one edge to 170 mm from the tip end. Centered about 75 mm from the tip end on the edge opposite the groove is a rounded protuberance about 25 mm in diameter.

Plate 106: Haft Fragment; HAY 159 (½X)

A second nearly complete haft from Kebarah Cave is designated here as KEB 41. It is somewhat smaller, the original length being about 230 mm, but the central portion of the shaft is missing, leaving only the tip and handle ends. It also has a somewhat stylized deer's head carved on its handle and a rounded protuberance on the edge opposite the groove, about 10 mm wide and centered about 45 mm from the tip end.

Both the above specimens were treated in the museum with a preservative coating, probably shellac, which obscured any fine scratches. As these are museum specimens of almost unique artistic value they could not of course be cleaned of this preservative. Three observations could be made however: (1) the handle area was slightly polished, evidently from extensive handling; (2) there was a slightly increased polish along the edges of the tool-shaft adjacent to the blade groove; (3) there was no wear or polish on or around

[15]Bone hooks interpreted as belt hooks have been found at the site of Çatal Hüyük (Mellaart 1967), Haçilar (Mellaart 1970), Erbaba (Bordaz 1970), and Nea Nikomedia (Rodden 1962).

the protuberance. It had seemed remotely possible that the protuberance had played a functional role, perhaps as a button around which a cord may have been fastened. As no wear was found to indicate that this was the case, it is far more likely that these protuberances were a stylistic element intended probably to give the appearance of a phallus. Another specimen, WAD 21, bears an obviously non-functional convex bump at the same point, which could be a further stylization of the rounded protuberances on KEB 40 and KEB 41.

In addition to these specimens the Natufian sample includes seventeen fragments. Nine include the tip of the haft, four are grooved shaft portions, and four include the haft handle. Only fragments showing a groove along one edge could confidently be considered a haft fragment. Of the shaft fragments WAD 29 retains the flint blades still in place, cemented by what appears to be a plant resin.

It had been hoped initially that scratches appearing on these fragments would provide a guide to the direction of movement of these flint knives when they were in use. Small patches of such scratches do occur on a number of specimens running transversely back from the grooved edge, but none of these is patently clear[16]. If these scratches were the result of tool-use they would indicate that the implement was used with a sweeping arm movement similar to that used with a modern sickle. This may have been possible were the tool used to cut reeds but would probably be ineffectual in reaping wild wheat, as the brittle-rachised heads of the grain would be knocked from the stalks by the force of the blow. Grain of necessity would be cut by holding the stalks in the hand and cutting through them with a sawing motion. Reaping of this kind has been observed in use in Kurdistan (R. S. Solecki, personal communication). In a recent study of wear-traces upon flint sickle blades, Unger-Hamilton (1989) notes that the striations found on Natufian sickle blades were most probably caused by abrasion from soil particles (as is generally the case with bone implements as well). She concludes that harvesting must have been done near the base of the stalks to account for the frequent scratches. This observation accords well with the patches of irregularly oriented scratches seen upon the sides of the sickle hafts.

The distribution of use-rounding and polish on the hafts' surfaces is of considerable interest. Most of the specimens (ORN 5, ORN 6, and WAD 29 are exceptions) show fairly heavy uniform polish on all protruding surfaces. Several specimens, however, (notably KEB 42, WAD 19, WAD 20, WAD 21) show greater rounding and polish along the edges which flank the burin-groove into which the blades were placed. No clear-cut axial scratches could be observed in this worn area but fine axial scratches are difficult to detect as they are obscured by the grain of the bone and the marks left by tool manufacture. Nevertheless, considering the general shape of the tool when

---

[16]One specimen, WAD 19, has very deep gouges running across the edges, both on the grooves and on the opposite edge. These could have nothing to do with the original function of the implement. Perhaps after breakage the haft was used to retouch a flake or flint implement.

the blades were in place, it is difficult to imagine this edge becoming worn in the way observed unless the tool was moved axially back-and-forth with a knife-like motion rather than a sweeping stroke[17].

Antler Tines

A number of cut-off or broken-off antler tines, mostly of red deer, have been found at Natufian sites. These specimens all have smoothed and polished surfaces and tips which are criss-crossed by many scratches. Consequently they have been preserved by the excavators as possible tools. It is likely that these tines did have some specific function as they had been intentionally collected by the Natufians. The tines of naturally occurring red deer antler, however, may be seen to be worn in a manner similar to these artifacts. The wear is generated during the lifetime of these animals and is probably due to their rubbing their antlers against the bark and branches of trees. If there is any additional wear on these specimens due to their later use it is obscured by these numerous natural scratches and polish so no further determination of the possible function of these artifacts can be made.

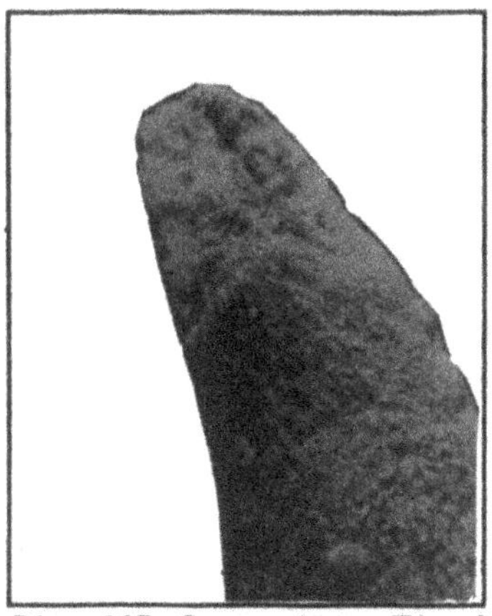

Plate 107: Scores Across Tip of Possible Flaker of Antler HAY 96 (2X)

Plate 108: HAY 96 (½X)

Possible flaker: HAY 96 is an exception. It is a large antler fragment (126.5 mm long) with an intentionally cut-off base, shows a number of deep score marks across the tip, which is also heavily chipped. This object may have served as a flaker for flint implements.

Flattened Implements ("Spatulas")

Two complete Natufian artifacts consist of flattened bone fragments. The first, HAY 197, is made on a large long bone fragment (140.7 mm long, 20.5 mm wide, 10.2 mm thick). Both ends of this elongated implement are rather broad and flat and irregularly rounded. The surface of the implement has been shaved to shape overall. All the high areas, including the sides and both ends are well rounded and highly polished. A large chip is missing from the smaller end. Only randomly-oriented scratches are visible on this implement.

---

[17] The evidence for the harvesting of wild grain by the Natufians is based primarily on the sickle gloss to be seen on many flint blades. It may be noted that on the one Natufian specimen available with flint blades in place there was no sickle-gloss visible on the blades. Some of the bone hafts may have been used for other purposes; a designation such as "blade-haft" would be more appropriate.

The second, RAK 3, is a large rib (216 mm long, 31 mm wide, 8 mm thick), which is largely unworked. There are a number of scores running axially on both inner and outer faces. One end appears slightly battered. This object is poorly preserved; no fine wear is visible.

Neither of these specimens has sufficiently unambiguous wear-patterns to provide much guidance to its function.

Toothed Object

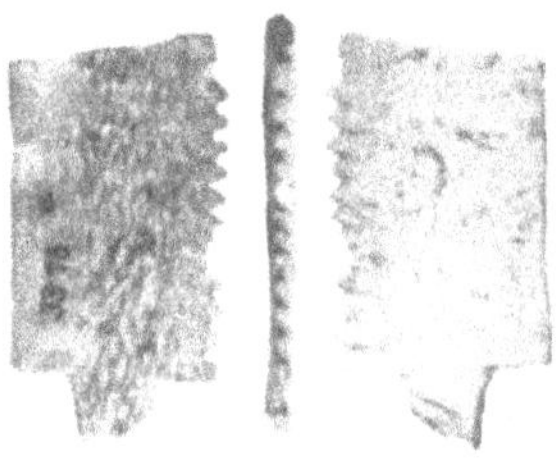

Plate 109: Toothed Object KEB 31 (½X)

A toothed object, KEB 31, bearing a row of short teeth along one edge, was found a Kebarah Cave. Turville-Petre considered this object a comb. It is broken at both ends; its present length is 41.6 mm, and it is 29.2 mm wide and 3.2 mm thick. The outer, slightly convex surface is decorated with a series of incised patterns. The object apparently originally had a narrower extension, almost surely a handle, extending from on end. This was broken off in antiquity, the break occurring at a weak point caused by a perforation which had been made at the base of the extension. A very small segment of the interior of this perforation remains on the specimen. Eleven notches about 2-3 mm deep were made along one edge of the object, evidently with a burin. These form twelve teeth (one broken off) about 2.5 mm wide at the "handle" end ranging to 1.5 mm wide at the other end. The edge opposite the notches is only slightly rounded and polished. This would not be surprising if the now missing extension were indeed a handle as this would have received most of the wear. The interior of the fragment of a perforation is quite well rounded and polished, indicating that a cord was passed through this hole, perhaps for suspension of the object.

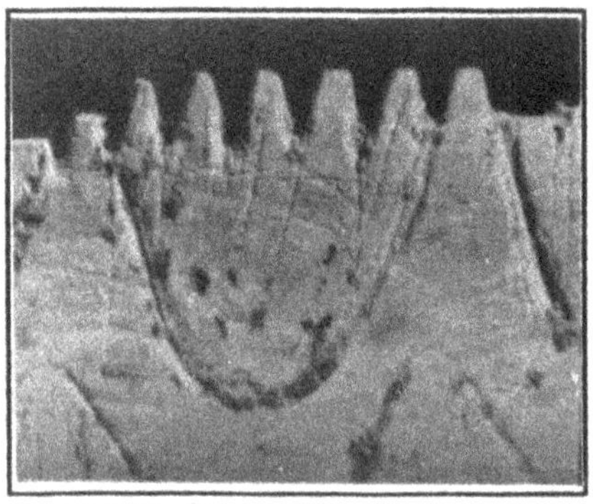

Plate 110: Incised Design on Toothed Object KEB 31 (2X)

Rounding and polish on the teeth, which is quite extensive, is confined to the tips and does not penetrate the bottoms of the notches. The bottoms of the notches are quite unworn. At 65 diameters magnification very fine scratches may be seen in the polish on the tooth tips, running transversely across the tip, at right angles to the object's axis. Some similar scratches may be seen within the notches near the tips of the teeth. No other wear is apparent on the specimen. This clearly indicates that the toothed edges were the active area. The scratches also indicate that the implement was moved in a back-and-forth motion, teeth down, at right angles to the tool's axis. The material worked penetrated slightly in order to account for the scratches on the sides of the teeth but it did not penetrate to the bases of the notches.

Whether an implement with such very short, blunt teeth would have functioned effectively as a hair comb is somewhat doubtful. Furthermore, hair would penetrate to the bases of the notches and cause wear there. This implement, however, has an outward form very much like a hair comb, especially if one considers its appearance with an extended handle and its fairly elaborate decoration. The wear also unambiguously indicates that it was moved in a manner very similar to that of a hair comb. An alternative hypothesis is that this implement might have served as a weaving tool, used to press down the weft. Both these hypotheses fail to account for the lack of wear at the bases of the notches. On balance one is led to consider Turville-Petre's original assessment as a bit more likely. The short teeth were probably the best that could be managed given the available flint implements and the broad teeth may have prevented the implement from penetrating sufficiently to result in wear at the base of the notches.

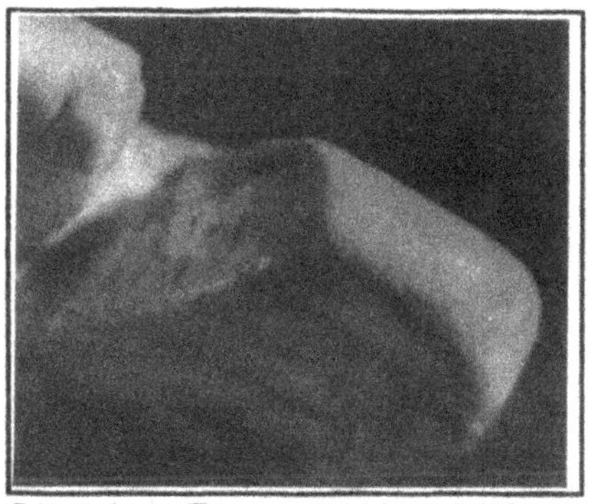

Plate 111: Toothed Object KEB 31; Rounding and Polish of Tooth Tip (10X)

Perforated Scapula (Probable Shaft Straightener)[18]

In 1928 C. Lambert recovered a perforated deer scapula from a trial trench in the cave of El Wad. This implement was briefly described by Garrod (1930) and attributed to the Natufian. In a later publication it is illustrated as belonging to El Wad Layer B2 (Lower Natufian)(Garrod 1932). Garrod remarks that the tool recalls the *baton-de-commandement* of the Western European Upper Paleolithic.

The implement consists of the left scapula of a deer from which most of the blade and spine have been broken. The remaining portion, from the articulation (glenoid cavity) to the broken-off end, measures 130 mm. A large hole (16 mm by 22 mm) is cut between the lateral and medial surfaces with its center 45 mm from the articular end. The interior of this hole are so heavily worn that only a few traces remain on one edge suggesting that the hole was bored by means of a large flint implement. The scapula is dark brown in color and lightly polished overall.

Observed wear

Raised portions at the distal end show some rounding and polish, probably from handling. Very marked wear is visible within the perforation. The entire interior surface of the hole is smoothed and polished and the exterior margin of the hole on both lateral and medial surfaces has been worn to a four-lobed quatrefoil shape. The lobes of the quatrefoil which lie on the axis of the tool are much larger (more worn) than those which lie at right angles

[18]This implement has been discussed in greater detail in a previous publication (Campana 1979).

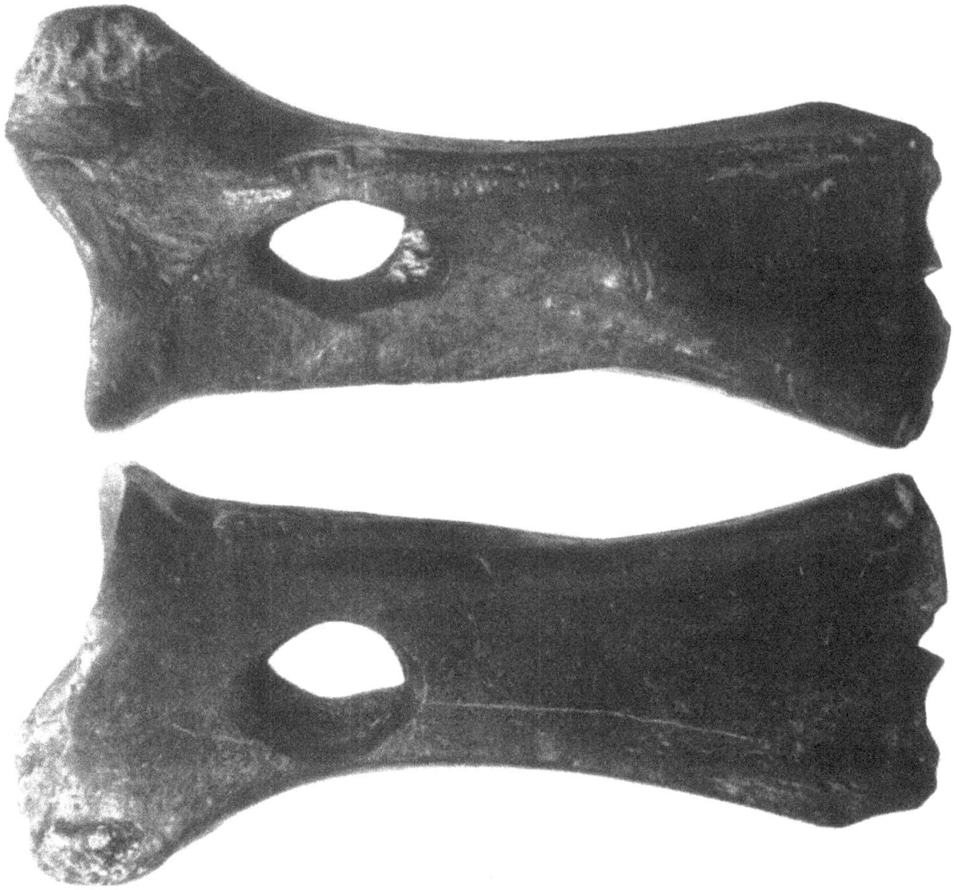

Plate 112: Perforated Scapula From El Wad; Probable Shaft Straightener (1X)

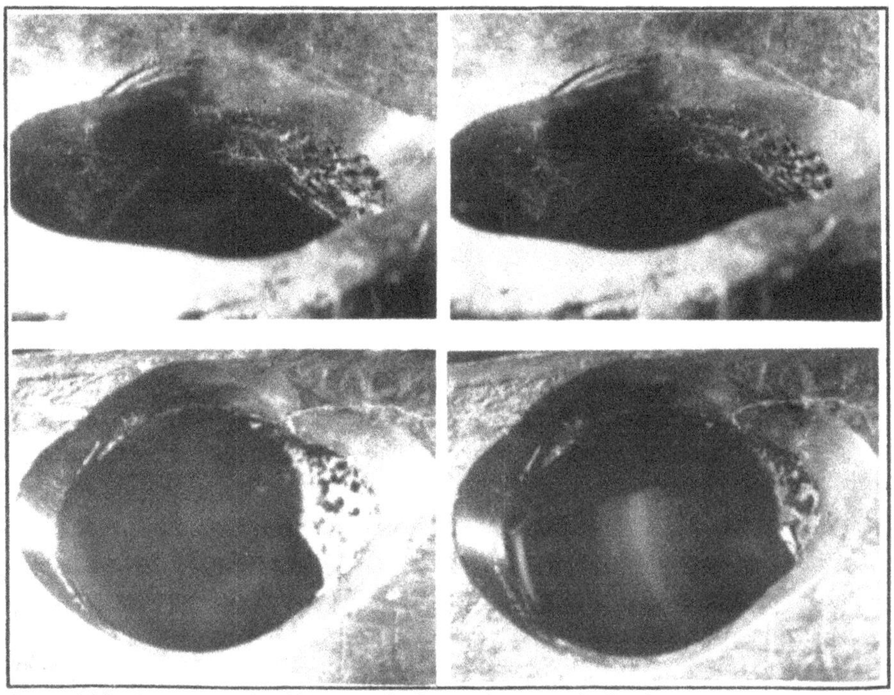

Plate 113: Stereoscopic Views of Both Sides of Perforation in El Wad Scapula (2X)

to the axis. The margins of the worn lobes are very distinct, forming an angle with the outer, unworn surface of the bone. There is very little rounding of these margins visible. Similarly, the intersection of the adjacent worn lobes is marked by a distinct angle or ridge with little visible rounding. These worn lobes are clearly cylindrical in shape and the surfaces of the worn lobes form angled cylindrical sections. Examined microscopically at 40X all the lobes show fine, parallel scratches running from the outer to the inner margin of the lobe and parallel to the axis of the cylinder.

Suggested uses

Among the possible uses suggested for this implement and others like it are (1) use as a thong stretcher or (2) use as a shaft wrench or straightener, or possibly as a shaft polisher. Given the observed wear-pattern the latter possibilities provide the most likely hypotheses. Shaft straighteners of essentially similar form are well known from ethnographic contexts. Jenness (1937) describes such a tool as used by the Eskimo. Similar tools were also used by the Navaho (Kluckhohn, Hill, and Kluckhohn 1971). These tools were made of bone or horn with single or multiple perforations. The shafts were straightened by running them back-and-forth in the hole while bending out the crooked portions. Perforated stone shaft straighteners as well as another type consisting of a grooved stone were used by the Miwok (Barrett and Gifford 1933).

Use of the El Wad tool as a shaft straightener would entail inserting the curved shaft into the hole and then twisting the implement so that torsion is applied to the shaft by the opposed lips of the hole. The implement must be moved up-and-down the shaft in order to straighten it evenly. Such an action would be expected to result in the observed wear-pattern. The shape of the worn lobes suggests that they were formed by friction with a cylindrical object. The lobes on alternate sides of the hole are approximately equal in depth and also in their angle to the bone surface. This would also be consistent with wear against a cylindrical rod angled in the hole. The abrupt, angular margins of the worn lobes indicate that this object was stiff like a wooden shaft rather than flexible like a leather thong, which would have resulted in rounded edges (see Chapter VII). Furthermore the direction of the fine scratches indicates that the shaft was moved back-and-forth rather than rotated. Finally, it was fairly easy to explain the shape of the worn hole if it functioned in the manner described. There would be two distinct ways to put torsion on a shaft inserted in the hole: (1) By grasping the tool in the manner of a wrench a maximum torsion may be put upon the shaft by taking advantage of the relatively long lever arm and the use of the whole arm. Held in this way the shaft would contact the edges of the hole on the distal edge of one surface and the proximal edge of the other surface. The relatively great torsion and the resultant friction would lead to the deep lobes observed. (2) By grasping the tool in the manner of a screwdriver a lesser degree of torsion may be placed upon the shaft, using the short lever arm formed by the width of the tool and the twisting of the wrist. The contact points would be at right angles to those of method (1). The lesser torsion and friction would be lead to relatively shallow worn lobes, as is observed. Intermediate positions of the tool would be difficult to achieve comfortably so that little wear would be expected between the lobes. This is as observed.

Experimental "shaft straightener"

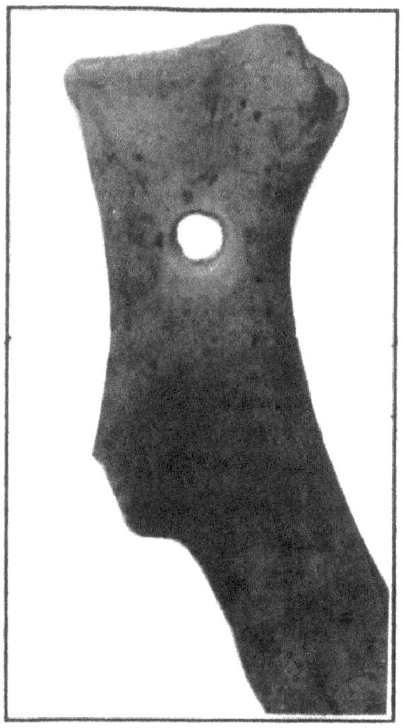

Plate 114: Experimental "Shaft Straightener" (½X)

To test the hypothesis that the wear-pattern observed on the artifact could have been the result of its use as a shaft straightener or polisher an experiment was performed. A cow scapula was pierced by means of a steel tool and the resultant hole reamed out using flint implement to simulate the ancient perforation made entirely by flint tools. The final hole was hourglass-shaped in cross-section and 12.6 mm in diameter at its narrowest point. A natural hardwood stick 10.3 mm in diameter was set in this hole diagonally to press on alternate edges. With the scapula held firmly in a clamp, this stick was then run back-and-forth through the hole by means of a reciprocating mechanical apparatus. This apparatus was constructed so as to have considerable play, allowing the stick to find its own position in the hole. In addition the apparatus was moved and readjusted numerous times during the experiment so as to simulate as closely as possible the wear that would have resulted had the experiment been done by hand. Sufficient torsional force was applied to the end of the stick to produce maximal friction without causing the stick to jam (estimated to be about 1 kg). Using a stroke length of 9 cm, and running it at a rate of twelve hundred strokes per hour, the stick was run through the perforation ten thousand times. As it was expected that there would have been a small amount of dirt present during the use of the artifact being simulated a few grains of fine sand were introduced at the beginning of the run, although no additional sand was added.

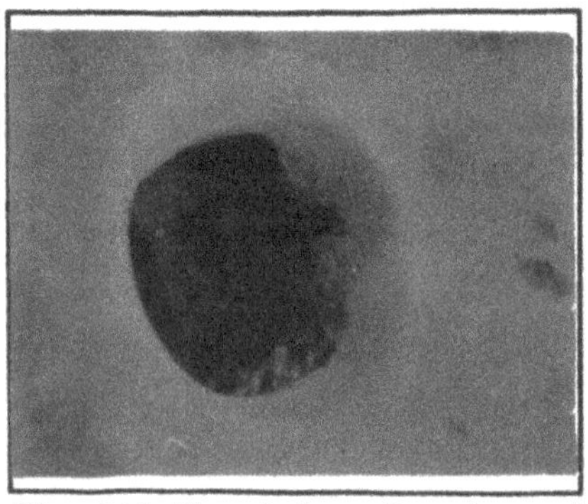

Plate 115: Worn Lobe on Experimental Tool (2X)

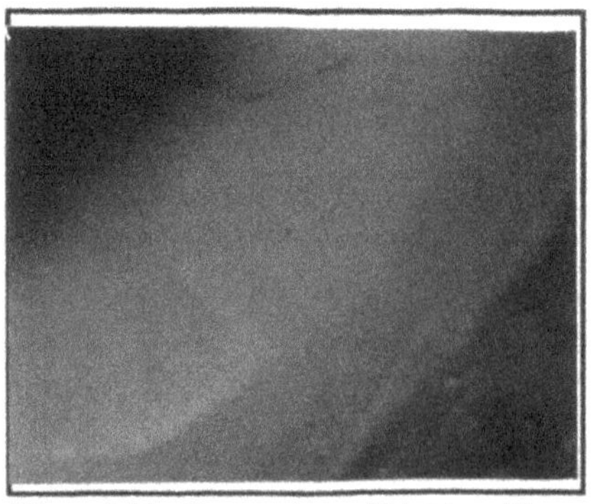

Plate 116: Scratches Within Worn Lobe of Experimental Tool (10X)

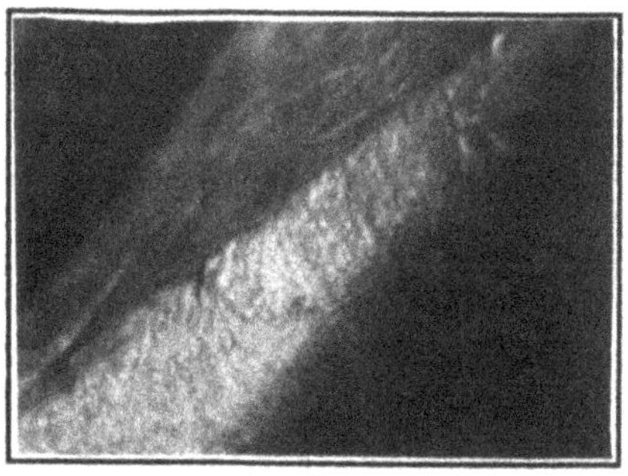

A

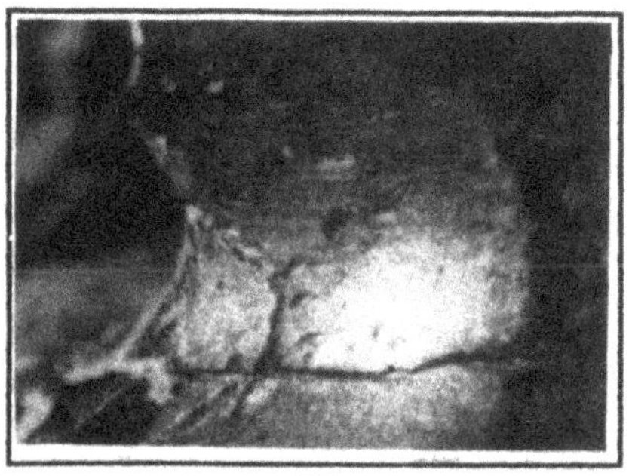

B

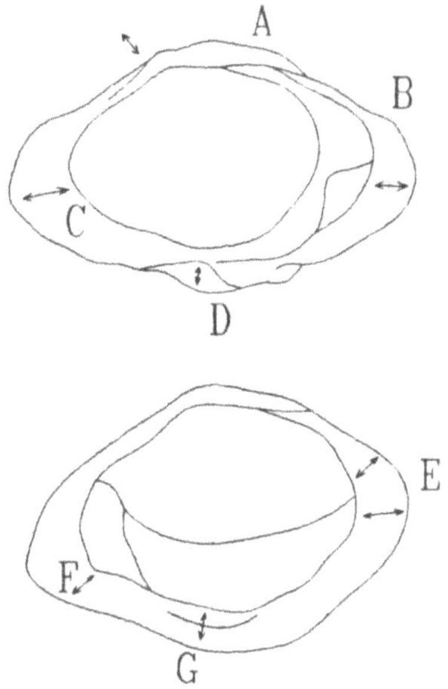

Plate 117: Locations of Distinctive Scratches on El Wad Perforated Scapula (Photographs 10X)

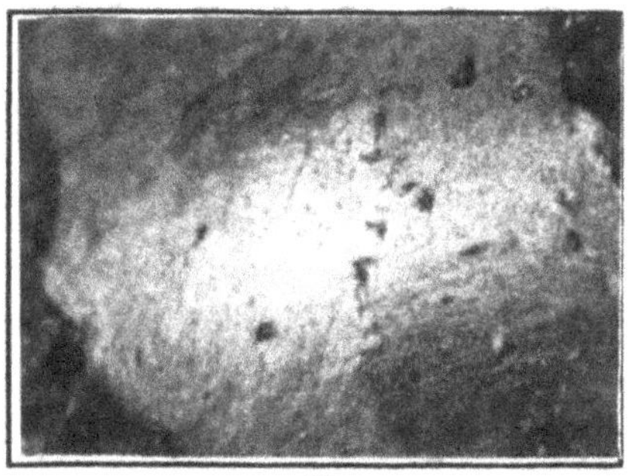

C

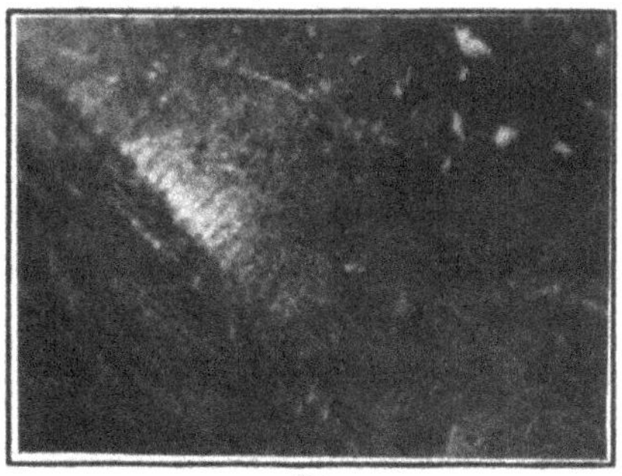

D

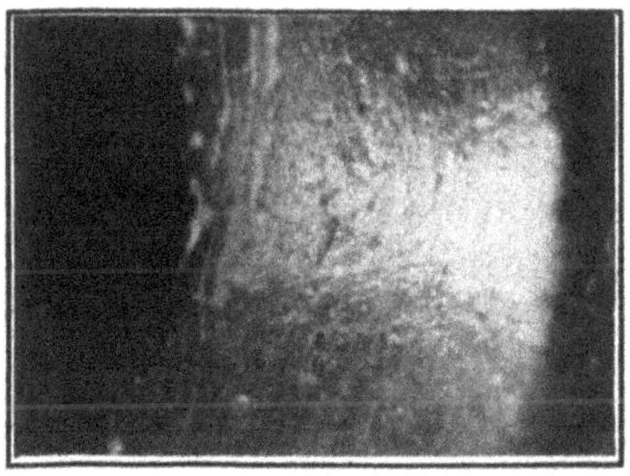

E

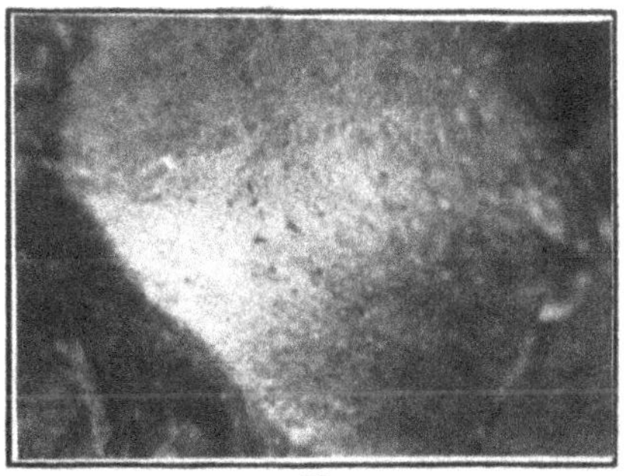

F

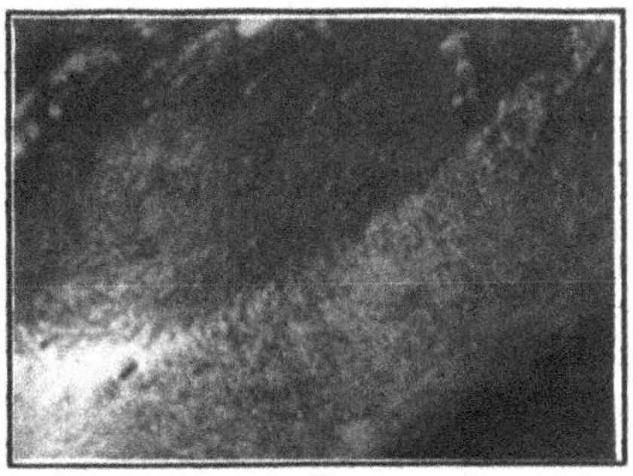

G

Comparison of wear-patterns on the El Wad implement and the experimental "shaft straightener"

The following similarities are noted:

1. Both the El Wad scapula and the experimental tool show semi-circular worn lobes on alternated edges of the perforation.

2. Both implements show distinct, angular rather than rounded margins on the worn lobes.

3. On both implements the cylinder-shaped worn-in lobes on alternate edges of the perforation are parallel.

4. Both implements show similarly placed fine parallel scratches running through the hole.

Given these similarities it may be taken as demonstrated that the El Wad implement probably functioned as a shaft straightener or polisher. (The wear-pattern of the thong stretcher will be shown in Chapter VII). Given this conclusion the following additional observations may be made:

5. The experimental tool shows a number of fairly coarse scratches, evidently caused by the presence of the few grains of sand. The great majority of the scratches are very fine on both the El Wad and the experimental tool, indicating that the tool was used in a fairly clean condition and without the addition of an abrasive. Therefore the El Wad specimen was not likely to have been used as an intentional smoothing implement. Some polish was observed on the stick used in the experiment; the El Wad tool could conceivably have served as a polisher, but not without bending the shaft as considerable torsion must be applied.

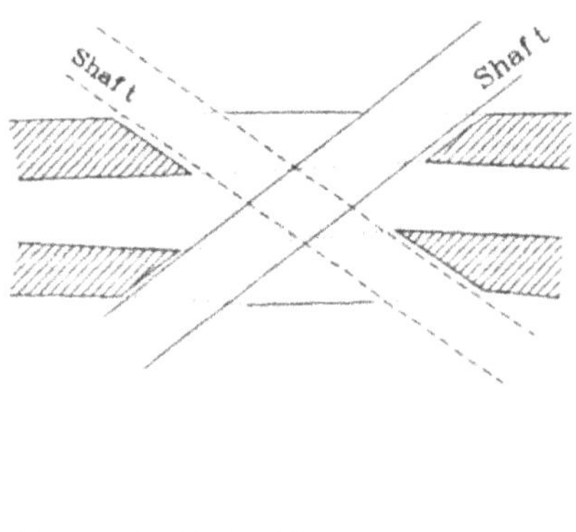

Figure 10: Method for Determining the Width of the Shaft Within the Worn Lobes of the Perforated Scapula

6. Observation of the experimental tool shows that the wooden shaft fits flat against the alternate lobes of the implement. The distance between the parallel bottoms of the worn lobes on alternate sides of the hole in the implement was found to be 10.3 mm, the diameter of the shaft which wore them. Applied to the El Wad artifact, as there are four pairs of well-defined lobes it is possible to make four such measurements. They are all equal to approximately 8.5 mm. Furthermore, a circle with a diameter of 8.5 mm was found to fit well within each of the worn lobes of the artifact.

Conclusion

The El Wad tool was probably used to straighten and/or polish a shaft approximately 8.5 mm in diameter, about the thickness of a pencil. The shaft may have been of wood or possibly reed. This is most likely to have been an arrow shaft, and probably one used for small game.

A final observation concerns the degree of wear visible on the El Wad tool. Eight hours of continuous wear produced a worn lobe of less than 0.5 mm depth on the experimental tool. The wear on the artifact is many times greater than this, the deepest of the lobes being about 6 mm deep. It is clear that this extensive wear represents a very protracted period of use. Evidently many shafts of about the same diameter were straightened by means of this tool.

Implement Handles

Several Natufian specimens appear to the worn handle ends of large flattened implements. They show rounding and polish on the edges and protrusions and randomly oriented scratches. KEB 28 is the handle end of an implement for which the functional end is missing. It is made on a large rib and is rounded at the base and sharply constricted in width toward the tip end. A perforation is near the base. There is considerable rounding and polishing on the sides, less on the faces, presumably from handling. The interior edges of the hole are rounded and polished, markedly more so on the basal side. KEB 78 is also a handle end somewhat similar to KEB 28 except that it is made on a large shaft fragment and is thicker. It also has a perforation near the base. The sides and base show rounding and polish from handling and the hole is slightly worn on the basal edge. It would appear from the wear in these holes that most of these implements were suspended from a thong.

CHAPTER VII

The Functions of Zagros Protoneolithic Bone Tools

Points and Pointed Implements

The Zagros Protoneolithic sample included 106 specimens which are generally similar to the Natufian specimens discussed in Chapter VI in having small, round tips. Of these, twenty are sufficiently complete to be considered complete implements; the remainder included here are tip fragments. The forms and dimensions of these specimens form a continuum which is difficult to subdivide. Some rather artificial divisions may be made however, and several specimens require individual description.

As is the case with the Natufian specimens, most of the complete Zagros Protoneolithic implements show rounding and polishing on the edges of the shaft and at the base indicative of prolonged handling. In general, this rounding and polish is not nearly as marked on the Zagros Protoneolithic specimens as on those from the Natufian, although there are exceptions.

Large implements:

Five specimens represent the largest implements in the collection. ZCS 57, somewhat fragmentary, is 109.5 mm long and is made of the unfused proximal radius of a small artiodactyl, with a bit of the shaft of the ulna attached. There is little rounding of the base and little polish, although this may be due to the conditions of preservation. The round tip is broken off but appears to have been quite large; the unworn break is 4.8 mm in diameter. Light rotation scratches surround the tip to a distance of 8 mm from the tip end. There are many fine scratches running transversely across the central area of the shaft.

ZCS 55 and ZCS 56 are similar tools, 124.0 mm and 142 mm long respectively. They are made on long bone shaft fragments. Both are rounded and polished on the shaft and base, clearly by handling. The tip of ZCS 55 is missing and the break is unworn. The tip of ZCS 56 is complete, 0.9 mm in diameter, and well rounded by use. Neither tool shows rotation scratches. ZCS 59 is a long narrow implement, 146.0 mm long, 12.0 mm wide. The tip is missing. Transverse scratches run across the shaft at the tip end. SHN 27 is a large wide implement, 84.0 mm long, 23.5 mm wide, rather crudely shaped on a shaft fragment, and with little rounding or polish. The large tip, 2.9 mm in diameter, is complete and slightly rounded and polished.

Intermediate-sized implements:

There are seven implements of intermediate size (80-95 mm long, 10.0-17.0 mm wide) formed from long bone shaft sections. These tools are in various stages of completeness. Tip sizes range from 0.4 to 1.8 mm. ZCS 62, ZCS 122, ZCS 128, and SHN 25 show light rotation scratches surrounding the tip. On ZCS 122 these extend to 30 mm back from the tip (7.5 mm wide at this point).

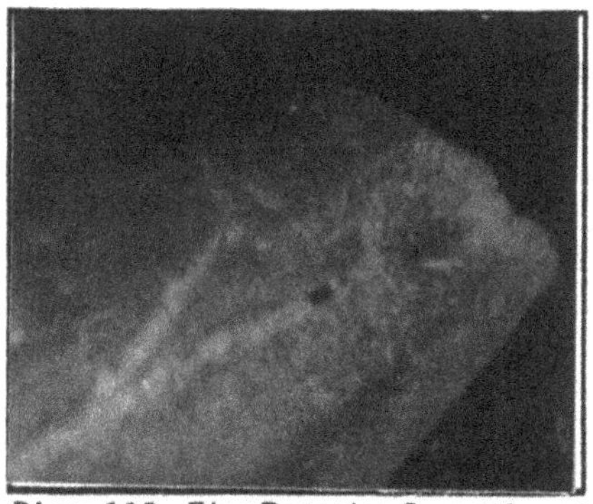

Plate 118: Fine Rotation Scratches at Tip of SHN 25 (34X)

Plate 119: SHN 25 (½X)

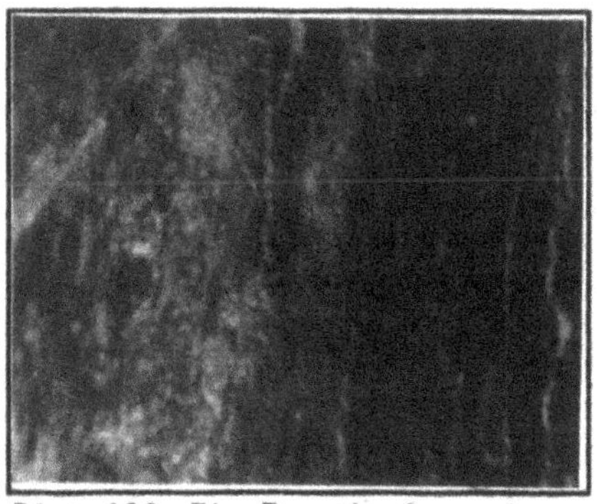

Plate 120: Fine Rotation Scratches on Shaft of ZCS 122 (34X)

Plate 121: ZCS 122 (½X)

Unique specimens:

Highly Polished Specimens: ZCS 15, which is a split long bone fragment showing a remaining trace of its articulation at its base, is heavily rounded and polished at the base and very highly rounded and polished on the central third of its shaft. It is considerably more highly polished on the flat faces than on the sides of the shaft. This is a relatively small object (51.5 mm long), missing its tip. The tip break, 3 mm in diameter, shows rounding indicating continued use. There are no rotation scratches visible but many randomly-oriented scratches are superimposed on the polished faces. It is possible that this implement served as a fastener and that the polish on the shaft was caused by friction with the fastened material, but the relative lack of polish on the sides of the shaft argues against this. The pronounced polish is more likely to be the result of prolonged handling. There is some slight polish extending only a few millimeters back from the tip, suggesting that the tip only slightly penetrated the worked material.

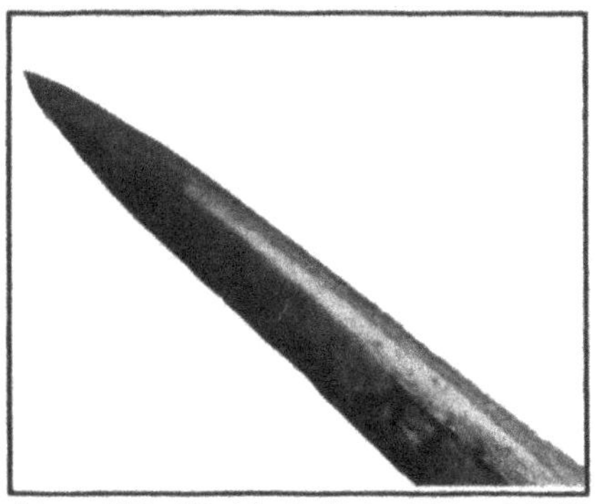

Plate 122: Highly Polished Tip of ZCS 15 (2X)

Plate 123: ZCS 15 (½X)

A larger specimen, ZCS 47, on a shaft fragment (91.0 mm long) has a somewhat similar pattern with marked polish on the outer face of the central third of the shaft. The tip is also missing, although the break shows some slight rounding, perhaps from continued use. There are no visible rotation scratches, but the tip end is highly rounded and polished, to about 35 mm back from the tip. This tool might possibly also have served as a fastener, but the greater wear on the face than on the edges suggests handling as the principle cause of wear on the upper shaft. The faces and sides of the tip third of the shaft are nearly parallel, the tool widening to approximately 7 mm at 35 mm from the tip, and the shaft has a nearly circular cross-section at this point. The area of polish extends the whole of this distance, and has surely been caused by pushing the tool into a soft material.

Implements with round shaft cross-sections: SHN 34 has a round shaft cross-section, and the faces and sides are nearly parallel all the way to the base. This is a narrow implement, only 5.2 mm wide. The tip is missing. The shaft and basal end are well-rounded and polished. Transverse scratches run across the shaft to 11 mm from the tip. ZCS 60 is similarly shaped, rounded and polished, but is missing both base and tip. ZCS 115 is also nearly round in cross-section for most of its length but widens at the its base. It shows relatively little rounding and polish and no clear scratches. Its tip is missing, but the break, 1.4 mm in diameter, is slightly rounded by use. Many of the broken fragments have round cross-sections, and may been broken from implements similar to these.

Miscellaneous Other Implements: SHN 28 is fashioned from a fragment of a scapula of a small artiodactyl. The unfused distal epiphysis forms the base and the tool shaft is formed at the intersection of the spine with the blade. The central area of the shaft is rounded and polished by handling, but the basal end is rounded to a lesser degree. The tip, somewhat ovoid in shape (3.8 mm by 1.5 mm) is complete and slightly rounded by use. There are no clear scratches.

ZCS 76 is a moderately large but very sturdy tool (82.0 mm long, 11.0 mm wide, 8.0 mm thick) formed on a shaft fragment. It is very dark in color. All raised areas on the shaft and base are heavily rounded and highly polished by handling. The conical portion of the shaft extending about 20 mm back from the tip is also highly polished. The tip has had two large opposed chips removed, leaving it with a wedge shape, 2.4 mm across. The edges of the tip are slightly rounded. Light rotation scratches surround the tip to a distance of 10 mm from the tip (6 mm wide at this point).

ZCS 29 differs from the other specimens in that it rather more fully worked and symmetrical. The shaft has an ovate cross-section and the sides and faces are smoothly curved. The tip, however, is round. The very tip is missing and shows no signs of continued use. There are some transverse scratches across the tip, but the principal indication of use is the more marked polishing of the shaft near the tip. There are several other fragments similar to this.

Some small fragments deserve mention. ZCS 18 is notable in that it is very narrow (47.3 mm long, 3.8 mm wide). Its shape is reminiscent of a heavy needle. The tip is round and complete, 0.6 mm in diameter. It shows moderate use-rounding. There are some randomly-oriented scratches on the tip and shaft, but no indication of rotation. There is a slight overall polish which may be due to conditions of deposition. ZCS 26 is unusual in that, while it is a very small object (22.8 mm long, 3.9 mm wide), the tip (0.8 mm in diameter) has been worn to a conical shape, evidently by rotation of the tool about its axis. The shaft near the tip is surrounded by heavy concentric rotation scratches. This would suggest rotation against a rather firm material, perhaps wood.

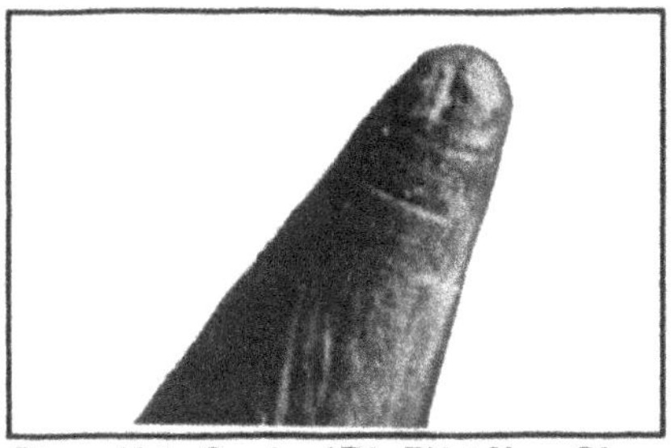

Plate 124: Conical Tip With Very Clear Rotation Scratches; ZCS 26 (10X)

Plate 125: ZCS 26 (1X)

Of the 106 specimens, sixty (56.6%) have unbroken tips. The diameter of these tips ranges from 0.4 mm to 3.8 mm (mean = 1.4 mm, s = 0.69). Of the sixty tips, twenty-one (35.0%) are greater than 1.5 mm in diameter, which appears to be the largest practical size for a perforator. If 1.5 mm is taken as the dividing line above which a tip may not be expected to easily penetrate hides, the complete tips may be divided as shown in Table 9.

TABLE 9

ZAGROS PROTONEOLITHIC IMPLEMENTS: RELATIONSHIP OF SCRATCHES TO TIP SIZE

	Below 1.5 mm	Above 1.5 mm	Total
Rotation scratches	16	6	22
No scratches	23	15	38
Total	39	21	60

NOTE: Corrected X^2 = 0.454, 1 d. f. Not significant at p = .05. Conclusion: both large and small tips are equally prone to show rotation scratches.

Overall, these scratches are found from 0.5 mm to 75 mm from the tip (mean = 10.7 mm, s = 15.2). The diameter of the shaft at this point ranges from 0.5 mm to 8.3 mm (mean = 3.5 mm, s = 2.3).

Conclusions

In terms of wear the Zagros Protoneolithic specimens can scarcely be differentiated from the Natufian sample taken as a whole. Further, the Zagros Protoneolithic implements cannot convincingly be sub-divided. The manner of use in general terms was clearly very similar to that of the

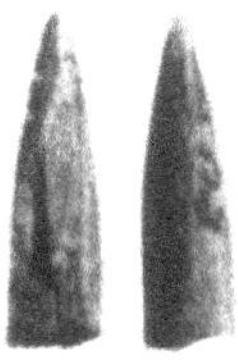

Plate 126: ZCS 43 (1X)

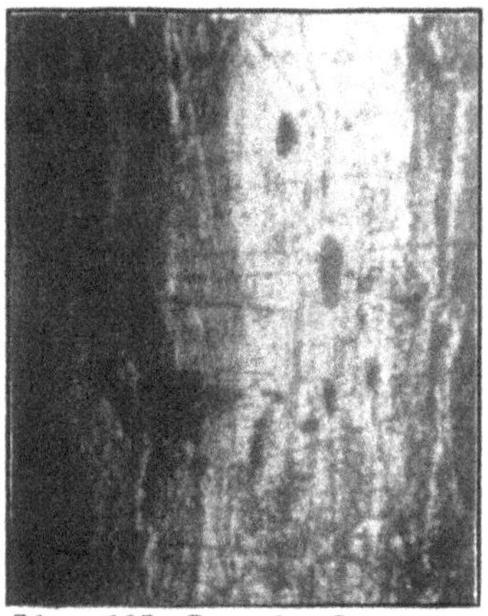

Plate 127: Rotation Scratches on Typical Tip Fragment ZCS 43 (67X)

Natufian tools: they were used to penetrate a fairly yielding material and this penetration was aided by twisting the tool about its axis. There is no indication, in the form of battering or shattering of the tips, that these implements served as projectile points, and this interpretation is contradicted by the presence of rotation scratches.

The major difficulty in the interpretation of these tools is the continuous gradation in tip size from small tips which could easily have penetrated hides to large tips which clearly could not. Experiment has shown that it is very easy to obtain very sharp tips using abrasive techniques, so that one is forced to conclude that these tips were intentionally made to such dimensions or at least that little attempt was made to resharpen them when they wore to such large sizes. There is no significant difference between the frequency of rotation scratches on large tips and on small tips.

It is likely that the larger tips were used to penetrate an open structure such as basketry, the smallest used as perforators for hides, and the intermediate sizes used for either or both purposes. In other words, there was probably a continuum of use similar to that suggested for some Natufian implements.

Spatulate Implements

The Zagros Protoneolithic assemblage includes a number of tool forms which may be loosely classified under the rubric of spatulate implement, as described previously.

Large, elliptical implement

Probable mattock or wedge: One Zagros Protoneolithic specimen is formally very similar to the Natufian specimen HAY 119. It is a tip fragment of antler, with a broad, spatulate tip and an elliptical profile. The tip width is 23.0 mm. This specimen is in a poor state of preservation and had to be pieced together from fragments. There is no use-rounding or polish visible. Although the tip edge is not chipped it appears somewhat battered. There are flattened areas on the base which could have been the result of striking with a hammer or mallet. The fine parallel scratches seen on the Natufian specimens do not appear on this tool; instead there are shorter, deeper scratches radiating from the edge of the tip. This tool might have been used as a mattock or quite possibly as a wedge, perhaps for splitting wood with the aid of a mallet.

Other Spatulate Implements

The Zagros Protoneolithic sites have produced a substantial number (26) of bone implements with spatulate tips. Most of these are incomplete; 16 are represented only by the very tip. Three are complete or nearly complete. The remaining six are larger shaft fragments which appear to lack only the base. In addition, SHN 17, which is a flat implement made from a rib, with a complete base and shaft but lacking a tip, will be described fully below.

If the tip width is compared to the blade width these implements may be divided into two distinct groups: (1) 14 implements for which the blade width nearly equals the total tip width (that is, roughly chisel shaped tips) and (2) nine specimens with blade widths substantially smaller than the total tip width (in effect, roughly elliptical tips). Two specimens, ZCS 77 and ZCS 87, have large fragments missing from the tips and could not be ascribed to one of these classes.

Broad-tipped specimens

These specimens range greatly in blade width from a minimum of 2.7 mm to a maximum of 11.0 mm, with a mean of 6.9 mm. This group includes two complete specimens.

Plate 128: Tip of Spatulate Tool ZCS 92 (2X)

Plate 129: Spatulate Tool; ZCS 92 (½X)

Probable hide dresser: ZCS 92 is made on the distal shaft end, including epiphysis, of a deer tibia, rather similar to those pointed implement on complete long bones previously described. The tip section is moderately broad (tip width = 9.0 mm, blade width = 7.5 mm). The high areas of the central shaft and the raised portions of the basal epiphysis are well rounded and highly polished, evidently by extensive handling. The tip edge of ZCS 92 is complete and undamaged by chipping. It is very smooth and highly polished to about 20 mm from the tip; the tip edge is smoothly rounded by use. This edge is quite thin and nearly sharp like a blade, although it does not actually form a cutting edge. The scratch pattern on this specimen is very marked. Leading back from the tip edge on both faces, but particularly on the inner face, are numerous very fine evenly spaced parallel scratches, running parallel to the tool axis. These are superimposed on the polish. Similar scratches also appear more than halfway up the shaft of the implement, superimposed on the rounded and polished area seen at the start of the diagonal break of the shaft made in forming the tip.

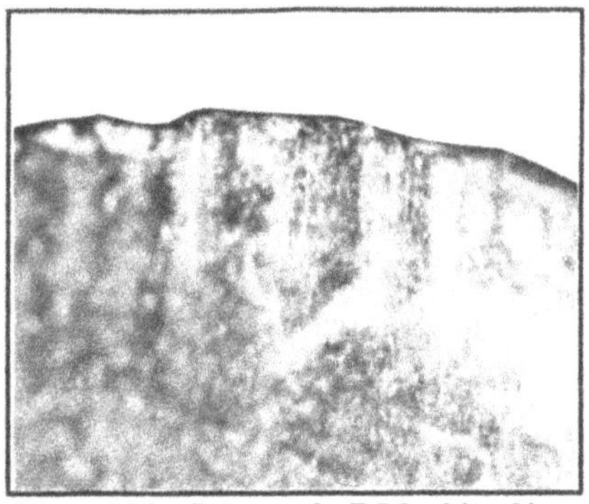

Plate 130: Edge of ZCS 92. Note Undulations and Distinct Fine Parallel Scratches (10X)

The most striking form of wear seen on this tool is the series of shallow, parallel grooves, apparently worn into the surface, running axially back from the tip edge for a distance of a few millimeters. These are fairly evenly spaced across the tip edge and have smoothly curved rounded profiles. The fine scratches are superimposed upon them.

It is evident that this tool was not subjected to very heavy use as a chisel, wedge or pry. The delicate edge would surely have been chipped and broken. The scratches clearly indicate that the tool was moved back-and-forth with rather long movements. The implement must have been held with the shaft at a very low angle to the worked surface to have resulted in the fine scratches seen at the base of the break on the shaft.

Two potential uses appear reasonable for this tool. (1) It could perhaps have been used as a wood-working implements, as a shaver or gouge. (2) It may have been used as a hide dresser, perhaps in the manner suggested for the Natufian specimens but more likely as a scraper or cleaner for removing the fat from the insides of hides, similar to the hide scrapers used by the Native North Americans. On the whole the wear seen on ZCS 92 seems more likely to have been the result of its use as a hide scraper than as a shaving or gouging implement for wood. Although both uses produce parallel axial scratches, those produced in shaving wood tend to be short and confined to the tip while those on ZCS 92 are fairly long. In addition, use of a bone tool as a shaver or gouge involves considerable pressure on the tool tip which tends to result in tip chipping and breakage; the specimen's tip is complete and unchipped. Although it has not been possible to reproduce the distinctive parallel grooves seen on the specimen, it seems more likely that these could have resulted from very prolonged rubbing against a soft, yielding material which could have flowed into the grooves and continued to wear them, even at their bottoms. A stiffer material such as wood would have tended to wear away the high points and eventually smoothed away irregularities rather than maintain or enhance them.

Possible hide dresser, later used as a chisel: The other complete specimen, ZCS 88, is made from a shaft fragment of a large long bone. It has a tip width of 12.5 mm and a blade width of 11.0 mm. Like the previous specimen the tip is well rounded and polished at the tip edge and shows numerous scratches running back axially from the tip. Unlike the previous specimen the tip edge is heavily chipped on both faces; the boundaries of these chips show little rounding. The central shaft is well rounded and polished, evidently by handling. The basal break, which is at right angles to the shaft axis, shows battering and flattening as if struck with a mallet. This implement seems to have been used in the manner of a chisel. The remnants of smooth polished surface and rounded edge seen between the chips does not accord with this use however. The parallel scratches at the tip are much like those of ZCS 92. The most likely interpretation is that this implement was originally used similarly to ZCS 92, but was later "mis-used" as a chisel.

Plate 131: Flattened Base of Spatulate Tool ZCS 88 (2X)

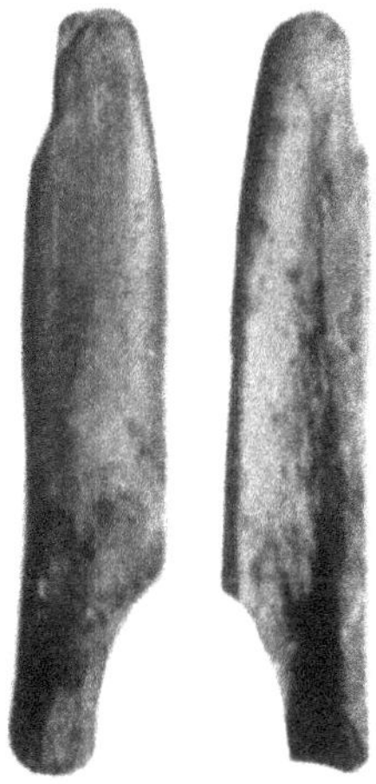

Plate 132: Spatulate Tool ZCS 88 (½X)

Conclusion: It is likely that all the broad-tipped specimens were used for the same purpose. Ten of the fourteen specimens assigned to this group show parallel or nearly-parallel fine scratches running axially back from the tip edge. It does not seem reasonable however to suggest that the narrower specimens could have been used as scrapers. A remaining possibility is that these tools may have been used in the tooling and decorating of leather. It may be noted that three specimens with fairly small blade widths, ZCS 84 with a blade width of 5.0 mm, ZCS 89 with a blade width of 10.0 mm, and ZCS 91 having a blade width of 7.0 mm also show the shallow, rounded parallel grooves at the tip noted for ZCS 92. None of these smaller implements possessed sufficiently sharp edges to have functioned as wood-working tools.

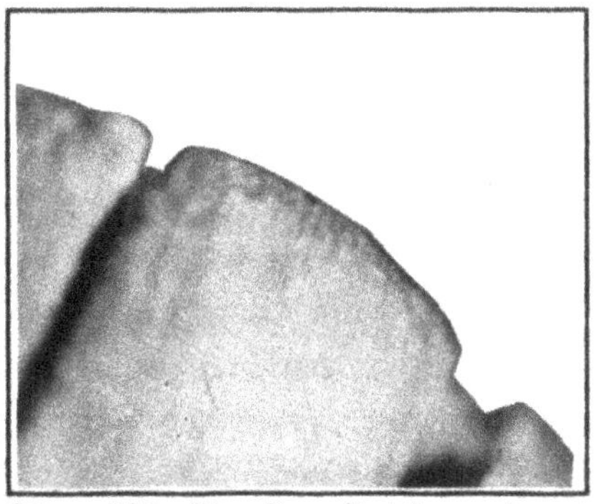

Plate 133: Fine Parallel Scratches at Tip of ZCS 89 (10X)

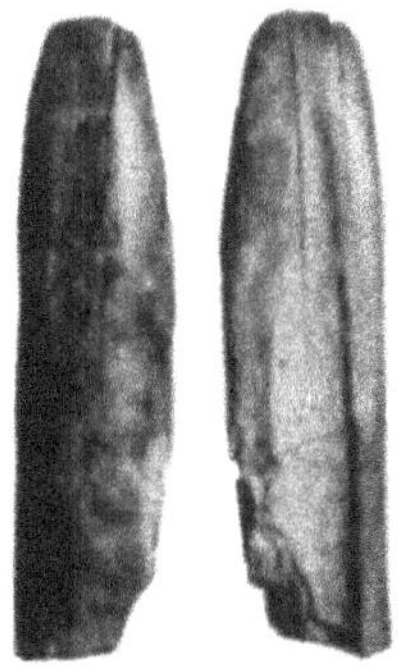

Plate 134: Spatulate Tool ZCS 89 (½X)

Elliptical-tipped specimens

This is a less homogeneous group, with considerable variation in tip shape.

Bodkin-like implements: The nearly complete specimen SHN 6 is included in this group. This is a long, fairly wide and flat implement (175.0 mm long, 22.0 mm wide, 6.0 mm thick). It is a very regular, fully worked, bodkin-shaped implement, with a flattened tip and basal perforation. The pattern of wear on this tool is nearly identical to that seen on the Natufian SHK 1, and it too almost surely served for manufacturing netting or mats. The tip is smoothly elliptical in profile. The tip edge is well rounded by use and is well polished at the tip. There are no chips on the tip edge. The polish is most prominent at the tip edges and is also most clear along the sides of the implement, being less marked on the faces. The polish on the sides may be the result of the tool having been held in the hand when in use. There are numerous scratches over the surface of the tool, but these are largely

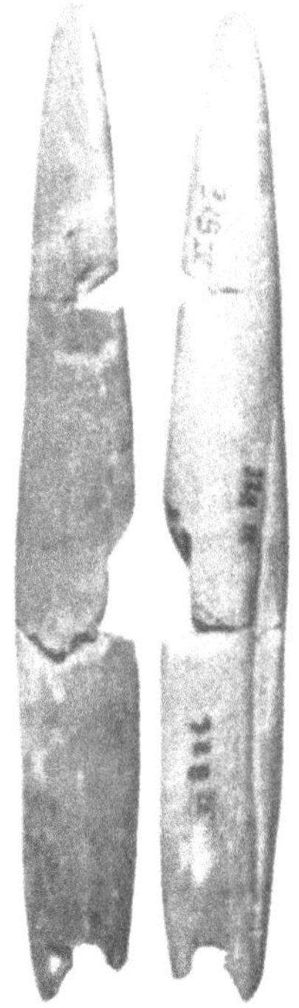

Plate 135: Bodkin-Like Tool SHN 6 (½X)

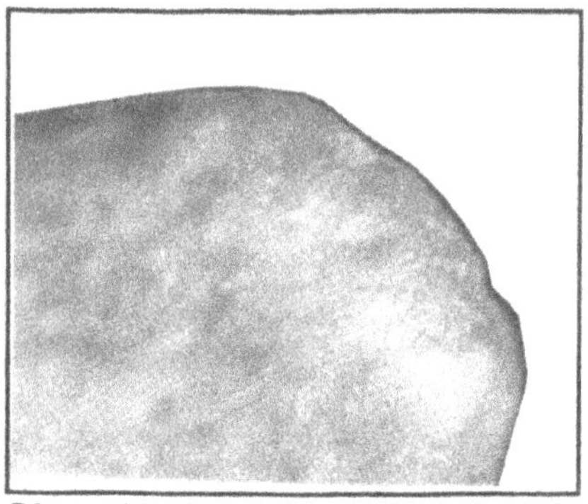

Plate 136: Scratches on Rounded Tip of SHN 6 (10X)

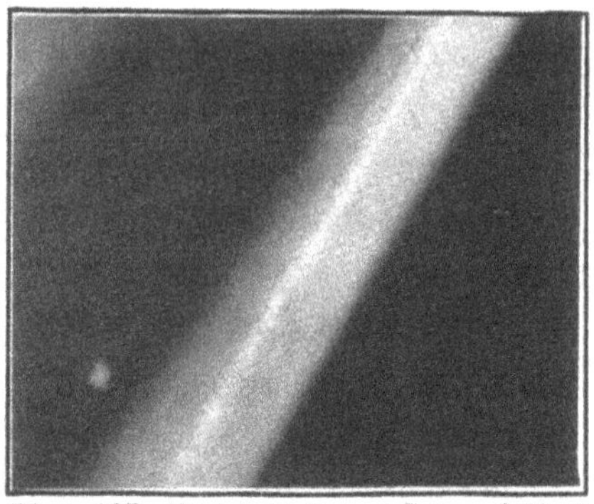

Plate 137: Highly Polished Edge of SHN 6 (10X)

Plate 138: Base of SHN 6 (1.4X)

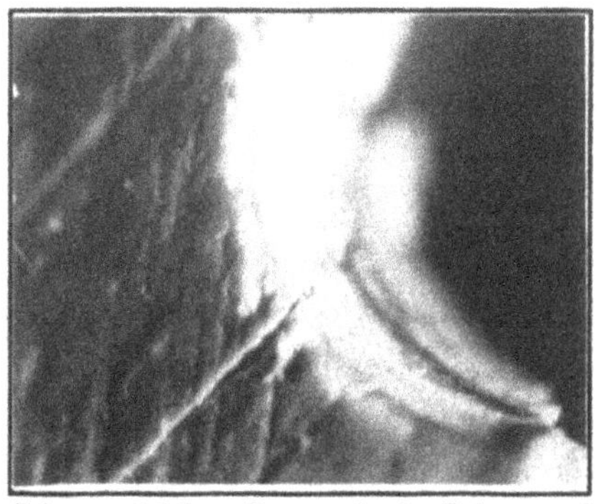

Plate 139: Rounded Edge of Perforation; SHN 6 (10X)

randomly oriented. At the tip fine scratches are visible leading back from the tip as well as numerous scratches crossing the faces. The latter are predominantly transverse. The hole at the base is 5.5 mm in diameter, thus limiting the size of the cord that might have been pulled by the implement. This perforation unfortunately is broken at the basal edge so that only the tip-side remains. The edges of the hole are slightly use-rounded and polished.

Six of the remaining specimens, all broken, could possibly be tip portions of similar implements. Four of them have scratch patterns at the tip generally similar to that of SHN 6. This pattern is not very distinctive, however, so it is not possible to assign these specimens with great confidence.

Plate 140: Bodkin-Like Tool SHN 17/18 (½X)

SHN 17/18 is a particularly fine bodkin-shaped specimen, generally like SHN 6 but made from a rib. The implement is decorated with cut-marks along the sides of the outer face. The tip is missing, but the sides show the same rounding and polish, probably from handling, seen on SHN 6. This specimen is particularly interesting in that it has two heavily worn perforations at the base. These perforations are very rounded and polished on the basal side and both show fine scratches running through the hole, from the friction of a cord.

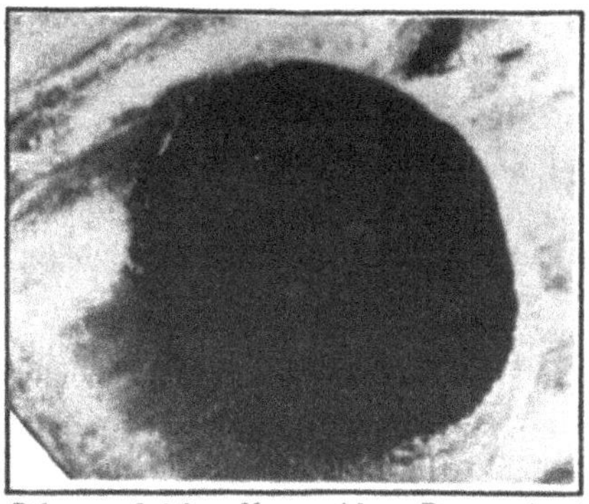

Plate 141: Heavily Rounded Perforation at Base of SHN 17/18 (10X)

Possible hide dressers: One specimen, ZCS 90, shows a worn chisel-like facet on one face of the tip. This area shows the rounded, parallel, axial grooves and fine parallel axial scratches common to the broad-bladed tools (interpreted as probable hide working implements) despite its elliptical tip shape. The two broken specimens which could not be classified on the basis of shape, ZCS 77 and ZCS 87, both show parallel axial scratches at the tip and were probably also similar in use to the broad-bladed implements.

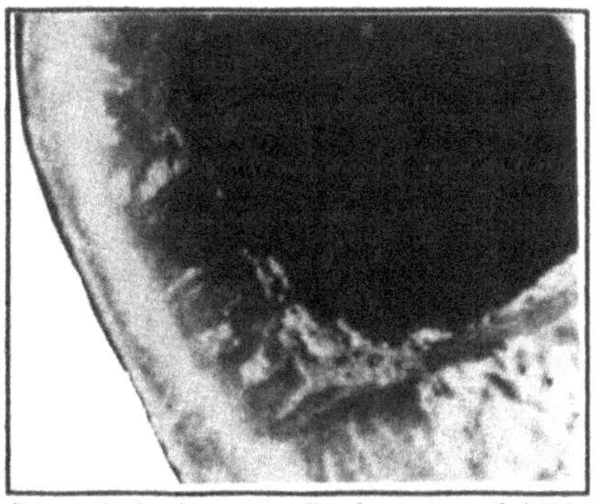

Plate 142: Opposite Perforation of SHN 17/18 (10X)

Bone Hafts

Shanidar Cave and Zawi Chem Shanidar produced two specimens (described fully by Solecki and Solecki 1963) resembling the Natufian ones in having a groove along one edge in which flint blades must have been mounted. On one of these (from Shanidar Cave) the single flint blade is still in place, cemented apparently with bitumen. As there was room for only this one short blade this is clearly a hafted knife, as identified by the excavators. The second specimen (from Zawi Chemi Shanidar) is shaped somewhat like a boomerang, elongated and symmetrical with a slight curve making the grooved edge slightly concave. Both ends are tapered. A V-shaped groove runs along the concave edge from one tip to the other. If this groove were filled with mounted flint blades, the implement must have been held at its broadened center. Held in this way the tool could only have been moved effectively back-and-forth axially, in the manner of a knife. Circumstances prevented the microscopic examination of the above Zagros Protoneolithic specimens.

Snapped-tipped Objects

Zawi Chemi Shanidar produced twenty-seven small bone points with snapped-off tips. These objects are quite variable in size, ranging in length for 7.6 mm to 43.5 mm, in width from 3.5 mm to 9.8 mm and in thickness form 2.5 mm to 5.5 mm. The size of the tip end at the break is much less variable, ranging from 1.2 mm to 3.1 mm. None of these objects shows any wear indicative of extensive handling, nor are there any visible scratches. Of twenty-eight tips only seven (25%) show some slight rounding of the snapped-off tip, and only one is slightly polished. This is a very low proportion in comparison to rounding and polishing appearing on points and pointed implements. All the others are apparently unworn. The only conclusion possible is that these objects were not intended as implements in their own

Plate 143: ZCS 138 (1X)

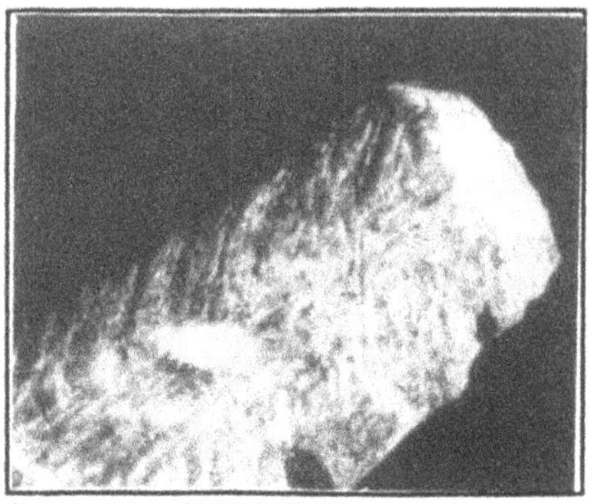

Plate 144: ZCS138; Unworn Snapped Tip (10X)

right, and only a few seem to have been used at all, perhaps for the same purposes as the points described above. As has been previously discussed, the most likely explanation for these objects is that they served as blanks for the preparation of small bone pins. These were broken off from the tip after shaping, leaving the snapped-tipped objects behind as waste. Two such pin-shaped objects are known, ZCS 4 and ZCS 14.

Plate 145: ZCS 141 (1X)

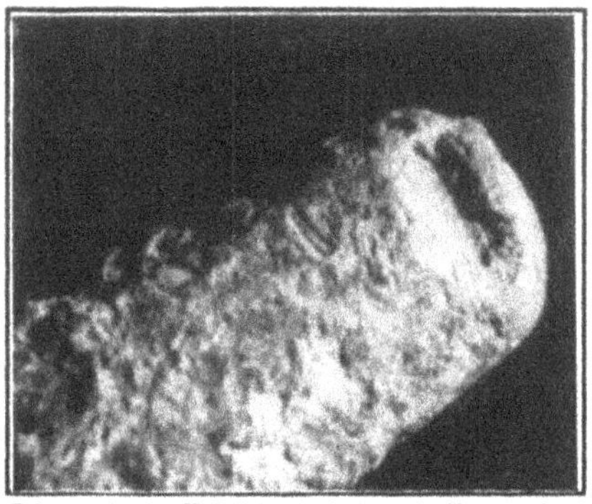

Plate 146: ZCS141; Slightly Rounded Tip (10X)

Related Pin-shaped Implements

ZCS 4 (18.0 mm long, 3.7 mm wide, 3.7 mm thick) has a rounded cross-section and one small tip (the very tip is broken off but the diameter of the tip break = 1.2 mm). The opposite basal end is rounded (3.7 mm diameter). Remnants of the score at which the object was broken off remain visible. The basal end is well rounded and polished and the edge of the break of the missing tip is slightly rounded. No clear scratches are visible.

Plate 147: ZCS 14 (1X)

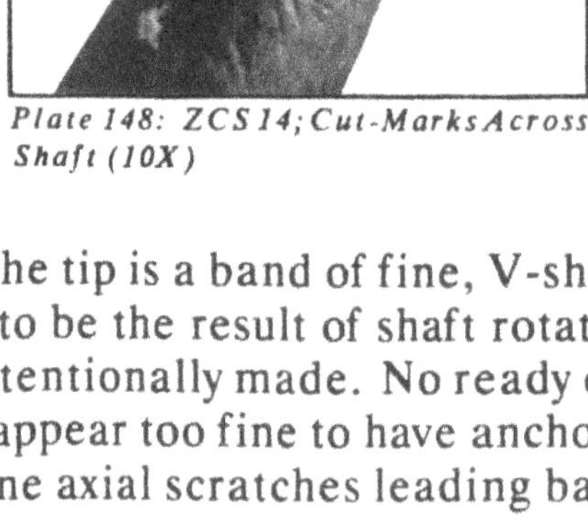

Plate 148: ZCS 14; Cut-Marks Across Shaft (10X)

ZCS 14 (16.7 mm long, 2.5 mm wide, 2.5 mm thick) is also round in cross-section. The very tip is missing but the broken end is very small (2.5 mm in diameter). This base was probably also scored and broken off but it appears to have been intentionally rounded. The base shows polish, probably from handling. The break at the tip is slightly rounded and the tip end is well polished. Surrounding the shaft about one-third of the way back from the tip is a band of fine, V-shaped cuts, all quite short. These do not appear to be the result of shaft rotation but rather short cuts which were probably intentionally made. No ready explanation can be offered for these cuts. They appear too fine to have anchored a cord. High magnification (100 X) shows fine axial scratches leading back from the tip.

Considering the very small, sharp tips and the extensive handling these specimens appear to have received they are likely to have been fasteners of some sort. They could have functioned as perforators but such small objects would not have been very convenient for such a use.

Antler Tines

There are a number of antler tines in the Zagros Protoneolithic assemblage. The use of these objects could not be determined for the reasons discussed in Chapter VI.

Flattened Implement

Plate 149: Coarse Scratches Across Face of Rib SHN 30 (10X)

A worked rib, SHN 30, was found at Shanidar Cave in several fragments. This implement was originally greater than 225 mm long. It shows considerable use-rounding and polish over its surface. This is most marked at the edges and particularly on the center section of the rib. There are numerous patches of fairly deep scratches running transversely across both inner and outer faces. This implement was probably moved back-and-forth in a direction roughly at right angles to its length. It was probably used as a rubber or scraper.

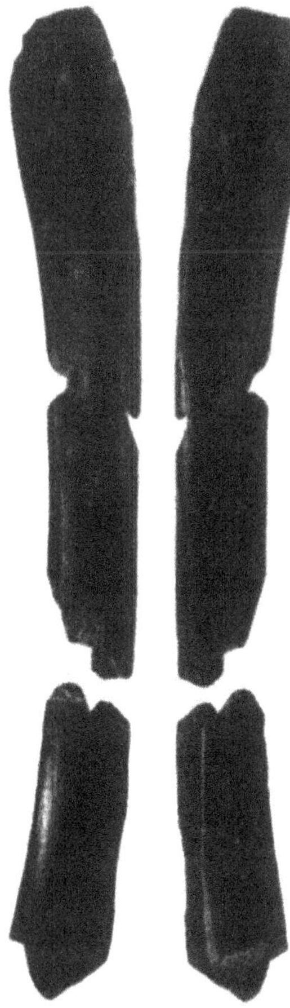

Plate 150: Rib; SHN 30 (½X)

Heavily Rounded Perforations ("Thong Stretchers" or Handles?)

Many specimens from both the Natufian and Zagros Protoneolithic assemblages have perforations bearing varying degrees of wear. The wear on these perforations usually takes the form of very marked rounding of the periphery of the hole, often accompanied by very fine scratches which pass through the perforation. It is clear that this wear is the result of friction from a cord or thong. Three possibilities present themselves:

1. In many cases the perforations are found at the base of the implement. In most cases the hole in this case would probably have served for suspension, with a thong or cord tied through it.

2. For the bodkin-like implement the perforation may have served to carry a cord, used in binding together mats or in making nets.

3. For some of the largest perforations, not obviously at the base of an implement of other type, it is possible that the perforation was in fact the active portion of the tool and that the implement may have served as a thong stretcher.

All these suggestions, it will be noted, require that a cord or thong pass through the hole to account for the observed wear. No other likely source of the observed wear can be suggested. However, to verify this suggestion and to gain some insight concerning wear rates the following experiment was performed (see also Campana, 1979).

Experimental "thong-stretcher"

Plate 151: Experimental "Thong Stretcher" (½X)

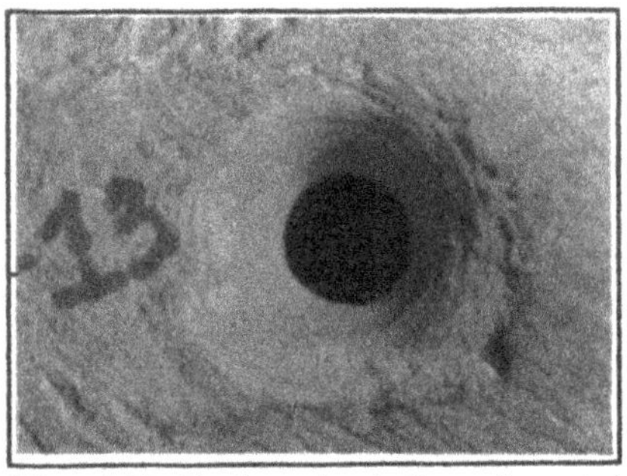

Plate 152: Perforation in Experimental "Thong Stretcher" (2X)

A fragment of cow long bone was perforated from both inner and outer sides by means of flint tools to produce an hourglass-shaped hole 6.0 mm in diameter, centering 15 mm from the end of the bone fragment. In order to test for the wear-patterns to be expected on a "thong-stretcher" a hand-cut leather thong 4 mm wide was passed though the hole and drawn backward over the end of the bone, so that the two ends of the thong formed an acute angle with each other. A weight of 250 g was placed on one end of the thong. The other end of the thong was then drawn back-and-forth through the hole, for a distance of 25 cm, by means of a mechanical device. This operation was repeated 10,000 times. During the course of this experiment the thong wore completely through three times and had to be replaced.

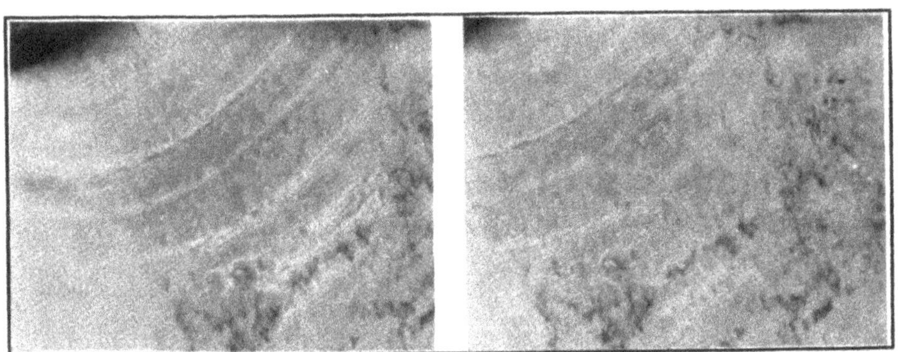

Plate 153: Stereogram of Rim of Perforation in Experimental "Thong Stretcher." Note Marked Rounding, Polish, and Fine Sub-Parallel Scratches Running Through Perforation (10X)

Comparison of wear on the perforated artifacts and on the experimental "thong-stretcher"

The following similarities and differences were observed:

1. The experimental tool shows a marked degree of rounding and polish on the interior of the perforation, particularly on the prominent ridge at the center of the "hourglass" of the cross-section. While several of the Natufian and Zagros Protoneolithic perforations have had nearly all traces of manufacture (concentric ridges within the hole) worn away on the basal side of the perforation, the wear on the experimental tool is comparatively slight and the traces of manufacture are still clearly visible.

2. Both the artifacts and the experimental tool show a marked worn-in lip on the outer surface, extending from the edge of the perforation toward the basal end. This is characterized by a smooth, even curvature without ridges and a glossy surface. The edges of the glossy patch are somewhat difficult to discern, rather than distinct.

3. There is virtually no wear visible on the inner cancellous surface of the experimental tool as the thong did not contact this area. On some of the artifacts, however, this area is very worn.

4. On both the artifacts and the experimental tools there are often extremely fine scratches (just visible at 40X) running through the hole and toward the basal end.

Conclusion

On the basis of this evidence it is clear that the observed wear-patterns within the perforations were caused by the friction of a cord or thong. The degree of wear on some of the artifacts, however, is so much greater than that seen on the experimental tool as to suggest a very prolonged period of use. This would be consistent with the use of the implement as a thong-stretcher, but the degree of rounding of the edges and the polish on the surfaces at the basal end must be taken into consideration. If, as is likely, this is the result of handling, such wear seems misplaced were the tool used as a thong-stretcher. A large amount of rounding from handling would also have taken a very long time to achieve. For most of the perforated artifacts it is most probable that the perforation served to hold a cord for suspension of the tool.

Miscellaneous objects

A few of the remaining miscellaneous specimens deserve some additional comment.

Vertebral spine with worn-in grooves: SHN 5 is a very unusual object from Shanidar Cave which is made on a fragment of a vertebral spine. A hole had been drilled through the broad, flat base. Across one of the wide faces

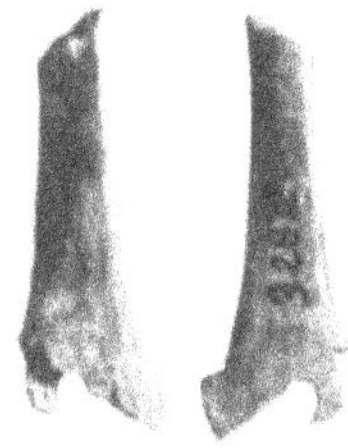

Plate 154: SHN 5 (½X)

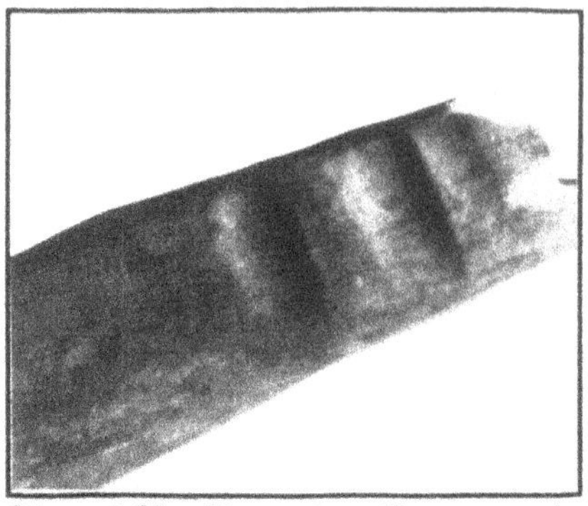

Plate 155: Worn-in Grooves in Mandible Fragment SHN 5 (2X)

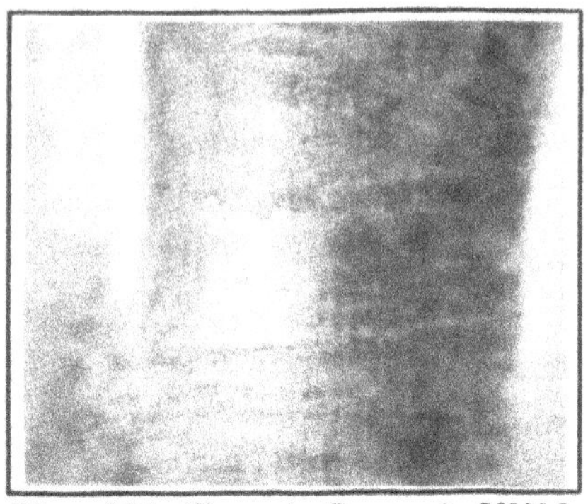

Plate 156: Worn-in Groove in SHN 5, Showing Fine Sub-Parallel Scratches (10X)

Plate 157: Fine Sub-Parallel Scratches in Groove of SHN 5 (67X)

of the shaft are three broad *worn-in* grooves. The edges of these grooves are all smoothly rounded and polished. Several similar but shallower depressions occur on the opposite face. High magnification reveals numerous sub-parallel fine scratches running through these grooves, transversely to the axis of the shaft. It is clear that the grooves are worn in by the friction of a cord. The interior edges of the hole are slightly rounded and polished as well. It is obvious that this specimen served as a bearing surface over which a cord was run at a considerable tension. The hole may have been used in hafting or lashing down the object. This specimen may have been used as a thong-stretcher, but its precise function cannot be determined.

Needle-like objects: ZCS 148 and ZCS 149 (see Plate 23) are two tiny bone objects, both fragmentary, which look like the eye portion of a needle. Both are round shafts, 3.5 mm in diameter, which are flattened and broadened toward one end. A hole (2.2 mm and 2.0 mm in diameter, respectively) was bored through the flattened section. Both specimens are broken across the hole. It is possible that the two objects are related but they do no fit together. There is very little wear in either perforation and little further interpretation can be made.

Other objects: A few other specimens appear to be decorative objects, waste fragments, and otherwise unidentifiable implement fragments. They will not be further discussed here[19].

[19] These objects are described briefly in Campana 1980.

CHAPTER VIII

Conclusions Regarding the Technology, Crafts, and Food-gathering Activities of the Natufian and Zagros Protoneolithic Peoples

Two types of evidence afforded by the study of bone tools lead us to conclusions concerning Natufian and Zagros Protoneolithic technology. The first is the marks left on the bone implements in the course of manufacture. They permit inferences concerning the techniques of manufacture used. The second is the forms of the implements and the use-wear appearing on them, which permit inferences concerning the tools' use.

The Manufacture of the Bone Tools

Quite unambiguous conclusions may be reached concerning the methods by which bone objects were made. The Natufian and Zagros Protoneolithic objects are strongly contrasted in this respect. In summary:

Basic Natufian manufacture methods

The Natufian bone objects, with only a few exceptions, were made by shaving and scraping bone fragments with a flint edge. Appropriate blanks for further working were frequently obtained by grooving bone shafts with a burin or other flint implement with a burin facet. Such grooves were used to split long bones, principally metapodials, into sections. Longitudinal grooves were made on opposite sides of the shaft and the bone shaft was then split by percussion. Circumferential cuts around bone shafts appear to have been made with a flint flake, perhaps denticulated, used in the manner of a saw. Small objects appear to have been made on thin bone blanks removed from the shaft of long bones by cutting two parallel grooves and removing the central splinter - the "groove and splinter" technique. All these manufacture techniques are similar to those used on the bone and antler objects of the European Upper Paleolithic and Mesolithic, which are also rich in bone objects. The "groove and splinter" technique specifically is well known in the European Mesolithic (Clark and Thompson 1954). Bone objects from the Levantine Upper Paleolithic and Kebaran periods (there are very few such objects) were also made by shaving bone with flint implements. The Natufian techniques applied to bone tool making are essentially ancient; it is the introduction of the bone implements themselves which represents the major technological change. The techniques used in making Natufian bone objects had probably been applied to the manufacture of wooden implements and objects since very early times. These techniques are not entirely appropriate to the working of bone, a much harder material. Grinding is a much more efficient method but was only very occasionally used for making tools, although it was more commonly applied to decorative bone objects.

In addition to edged flint implements and burins the Natufians made use of some form of device which allowed them to drill round, straight-sided holes. This apparatus surely provided a means for holding a drill shaft vertically and rotating it smoothly. It may have taken the form of a pump drill or bow drill, but this, of course, cannot be determined. It was surely tipped with a flint bit. Most of the perforations, however, were clearly made with a hand-held flint borer, so drilling devices could not have been widespread.

The Natufians rarely made any effort to smooth or polish bone objects, although these objects often became highly polished from later handling. The disposition of this polish on high areas and on parts which were clearly those which were gripped indicates that this polish was not intentional.

Basic Zagros Protoneolithic manufacture methods

The Zagros Protoneolithic bone objects, in contrast to those of the Natufian, were almost entirely manufactured using abrasive techniques. These objects were ground to shape against an abrasive surface with a moderately fine grit, probably sandstone or a similar material. This work was probably done using water as a lubricant and to flush the stone of waste. Round objects and those with sharp points were ground in a direction parallel to the tool axis, resulting in symmetrical, regularly-shaped objects. Implements with broad, flattened tips were ground with a motion transverse to the tool's axis.

Implements with burin-like facets appear to have been used to groove and section bones in a manner very similar to that of the Natufian assemblage, but examples are far fewer. There is no clear evidence for the "groove and splinter" technique. As in the Natufian, however, incised grooves were used to decorate the surface of some bone objects.

As in the Natufian, a few objects with round, straight-sided holes suggest the occasional use of some form of drilling apparatus. Most holes, however, were clearly made with hand-held flint borers.

While the Natufian bone tool manufacture indicates continuity with the Upper Paleolithic, in the Zagros Protoneolithic there is a clear break. Abrasive techniques were widely used at Zagros Protoneolithic sites; stone implements with ground edges were also present in the later phases. As stated above, abrasive techniques are more efficient than shaving for shaping hard materials such as bone (although they place some constraints upon the forms of bone objects possible to fabricate). The use of this method on stone implements is a characteristic technology of the Neolithic. Today the Neolithic is defined in terms of a constellation of cultural features, chief of which is the domestication of plants and animals. The introduction of abrasive techniques, however, was an important technological development. Ground stone and bone implements could be made with very regular contours and very sharp edges. This is of more than aesthetic importance. Such tools are more efficient in use, as they are easily resharpened, and thus far longer lasting. This represents a saving both in raw materials and in effort in fabrication.

Although the Natufian and Zagros Protoneolithic bone tool assemblages have simple forms in common, the wide disparity of manufacture techniques indicates that there is little affinity between the two assemblages.

The Functions of the Bone Tools

The forms of the bone implements themselves and the use-wear appearing on them provide evidence of the craft and food-getting activities of the Natufian and Zagros Protoneolithic peoples. This evidence is inevitably less positive than that.regarding manufacture techniques, but it is very suggestive when the assemblages as a whole are considered. These assemblages and the activities they suggest will be briefly reviewed.

Craft-related activities

In both the Natufian and Zagros Protoneolithic bone tool assemblages by far the largest groups are sharp-tipped points and pointed implements. These implements form about 52% of the Natufian sample and about 56% of the Zagros Protoneolithic sample. The Natufian sample of pointed implements can be sub-divided into several sub-groups based upon form. The Zagros Protoneolithic sample is largely fragmentary and cannot be easily sub-divided. It is highly probable that these difference in form reflect differences in function, and this is surely so between implements of widely different size. In terms of wear, however, these implements are all very similar. Almost all show some degree of rounding of the sharp edges of the tip and polish localized to the tip area. A significant proportion show fine scratches surrounding the tip. Most of the complete specimens show rounding and polish on the high areas of the base and shaft indicating that these tools were held in the hand during use. The wear indicates that these tools were pressed through a yielding material, often with a twisting motion to aid penetration. It is quite clear that the great majority were used as perforators or awls. It is more difficult to determine the nature of the material penetrated. Many of these tools may have been used in the working of hides or leather. Experiment has shown that the bone points must be kept quite sharp to perforate leather effectively. Points used for basketry need not be so sharp - indeed slightly rounded points are probably superior for this use. If an effort has been made to keep tips as sharp as possible, and slightly larger broken tips show little sign of continued use, hide perforation rather than basketry is indicated.

The Natufian assemblies include several large implements, mostly of antler, with broad, smooth, uniformly curved wear surfaces. These surfaces are well polished from friction with a soft, yielding material, and bear fine, closely-spaced parallel axial scratches which often run the length of the worn surface. These scratches indicate that these tools were moved with long back-and-forth axial strokes. They appear to have been rubbers for hide preparation, used to compress, thin, and smooth the leather surface, and generally to improve the suppleness and quality of the material. A tiny trace of red pigment found impregnating the tip of one of these tools suggests that this substance may have been used to color the surface of the leather.

There is only one large implement in the Zagros Protoneolithic sample similar to those discussed above. Instead, the Zagros Protoneolithic sample includes a substantial number of bone implements with broad, rather chisel-shaped tips. The working surfaces of all these tools are much smaller than those of the large Natufian tools; the wear-pattern, however, is quite similar. Most show closely-spaced very fine parallel axial scratches leading back from the tip edge. The exact function of these tools is uncertain; the broader-bladed tools may have been used as hide scrapers but it seems more likely that the small-bladed specimens would have been used in the tooling and decorating of leather. These tools are all similarly worn, and almost surely used for the same or very similar purposes. The difference in blade width is very great however; too great to be unintentional. This implies that these tools form a "set" in which the implements of different sizes were used for specific but related tasks, as do the different sizes of blade in a set of chisels or screwdrivers. Regardless of the exact function which they served, the existence of this set of tools is suggestive. It implies that the use to which they were put was carried out with a high degree of craftsmanship.

In both cultures the working of hides must have been major activities. The working of hide was surely a very ancient craft, and bone points, probably also perforators, were made throughout the Upper Paleolithic. These bone implements were few in number, however. What is remarkable is the great expansion in absolute numbers of these artifacts which occurs in the Natufian and Zagros Protoneolithic periods. These more specialized tools also imply a correspondingly higher level of craftsmanship.

Both the Natufian and Zagros Protoneolithic assemblages of points and pointed implements contain numerous specimens with tips too large to have readily penetrated skins or leather. It is highly probable that these tools were used in the manufacture of basketry. Such tools could have been used as manipulators to insert and twine together the basketry material. This would account satisfactorily for the patterns of scratches found high up on the shafts of some of the larger tools. Basketry was certainly made by the Zagros Protoneolithic people, as a small fragment of basketry has been recovered. No such direct evidence has yet been obtained for the Natufian culture. However, the wide variety of pointed bone tools suitable for the purpose, ranging from the relatively small specimens on split distal metapodials to the large implements on complete bone shafts, suggests that basketry of similarly wide variety may have been made.

Basketry, matting, net-making, weaving, and similar textile arts are closely related crafts which share many techniques and among which no clear dividing lines can be established. Simple implements such as awls can equally well be used for mat-making and basketry, and it is impossible to distinguish these similar functions on the basis of form or wear. Weaving, however, often calls for more specialized tools specific to this craft. There is ample evidence for weaving in both the Natufian and Zagros Protoneolithic cultures if that craft is broadly defined to include the making of coarse-plaited materials such as matting. Tools that seem to be associated weaving make up about 6% of both the Natufian and Zagros Protoneolithic bone samples.

Both assemblages include objects which were almost surely bodkins[20]. These are rather large, elongated implements with a flat cross-section. The flattened tip is smoothly curved. A perforation is usually made through the basal end. Some of these tools are made from ribs, others from bone shaft sections which have been flattened. The wear appearing in similarly shaped Natufian and Zagros Protoneolithic tools is not particularly distinctive, being confined to rounding and polishing of the tip portion and along the sides of the implement's shaft. Such scratches as appear are randomly placed and oriented, but there is a predominance of transverse scratches across the faces near the tips. The perforation at the base is often heavily worn on the edge nearest the base indicating the friction of a cord. While this evidence does not provide a positive indication of the direction of tool motion it does not contradict the interpretation of these tools as bodkins, and does not differ greatly from that observed on ethnographic specimens used for mat making[21].

In shape these tools do not differ substantially from modern bodkins used in non-mechanical weaving techniques such as tapestry-making. Such bodkins are used in conjunction with a very simple loom, often without a heddle. A plaited weave is made by passing the bodkin alternatively over and under adjacent threads of the warp. As the Natufian and Zagros Protoneolithic specimens are fairly thick such a fabric would necessarily have a relatively wide-spaced warp to allow the passage between alternate threads. The perforation of the bodkin-like implements provides an upper limit to the thickness of the cord or yarn used for the weft. As these perforations are of the order of 2-3 mm in diameter the cord or yarn must have been thinner than this, perhaps considerably thinner. It would have been possible to weave a yarn of such dimensions into a coarse but usable fabric[22]. If so it would imply

---

[20] Some suggestive corroborative evidence, apart from that of the bone implements, exists for weaving by the Natufians. A number of bone objects are decorated with a fine incised pattern that closely resembles a plaited fabric.

[21] A number of Ojibwa mat-making needles, housed at the Museum of the American Indian, New York, were examined for comparative purposes.

[22] An unusual example from Shanidar Cave has two perforations at the base, both of which are well worn. This suggests that this tool carried two wefts simultaneously. This implement is probably somewhat later in date.

that this fabric was at least moderately closely spaced. The fiber used for this cord or yarn is unknown, but animal hair would have been readily available. Alternatively the woven material may have been matting, for which rushes or reeds or similar material would have served for the warp. A weft of thinner cord is plaited or twined between the warp. It does not seem likely that these two possible uses could be distinguished on the basis of wear appearing on the bodkin-like tools but it should be emphasized that they are technologically very similar.

Despite minor difference in the implements, the Natufian and Zagros Protoneolithic specimens are sufficiently alike to suggest that weaving was at much the same state of development in both cultures.

Hunting and fishing

For both the Natufian and Zagros Protoneolithic cultures bone implements served primarily for craft activities and only a very few were of direct use in hunting, fishing, and gathering. In the Natufian sample of 312 objects only 17 barbed points were likely used for hunting. The barbed points could scarcely have functioned as other than projectile points. They usually show chipping and shattering of the tip of a sort to be expected from impact with a hard surface. The barbs often show considerable rounding and polish, suggesting long, repetitive use. With one exception these objects are made from bone, rather than antler as was common in the European Upper Paleolithic. Most are quite small. They have a single row of barbs and the complete specimens have a sharpened base. Several specimens show notable rounding and polishing of this basal tip, sometimes accompanied by fine rotation scratches around the shaft. This suggests that these barbed points were not permanently hafted but may have been detachable in the manner of a harpoon. If this were the case these points must have been attached to a retaining line, but no evidence for this was found.

It is very questionable that the small double-points ("gorgets") were used in fishing. According to this hypothesis the implement would have been secured by a cord around its center. On being swallowed by a fish it would have rotated and wedged itself in the fish's mouth. The evidence concerning these objects is to some extent contradictory. The majority of the specimens show no evidence that a cord was secured at the center, but a few have incisions across the center which could have been meant to hold a cord (these do not show any wear). A significant proportion of these tools, however, show fine-to-deep rotation scratches surrounding the tips. It would be difficult to account for these scratches if the small double-points were used for fishing. The scratches clearly indicated that these tools were forced into some material with a twisting motion, but the precise use remains unknown. It should be noted that there is a higher proportion of deep scratches on these objects then on any other pointed implements. The most likely use may be as simple fasteners or toggles.

The existence of bone hooks is often cited as evidence for Natufian fishing. There are, in fact, only four specimens known, of which two are fragmentary[23]. The wear on these few specimens is ambiguous. The tips of these hooks show very little wear, but they could not be expected to be as heavily worn as would t he tips of perforators. The most prominent wear is some slight rounding of the edges at the bottom of the "U" of these hooks. This may be attributable to extensive handling. Two specimens show notches around one leg which very probably held a cord; in one instance this cord may have been cemented in place.

Perhaps the most likely fishing implement is the single large barbed point made from antler. This object has a wide base, and was almost surely hafted. The shaft is slightly curved, and the barbs farthest from the tip are notably more worn than those near the tip or the tip itself. This implement is reminiscent of leister prongs (often found in pairs) known from Maglemosian sites (Clark 1953: 122). On a leister the barbed points are hafted at the end of a spear with their barbs facing one another. The fish is captured by wedging it between the sets of barbs. The wear appearing on this tool would be consistent with such a use.

The gathering of wild grain by the Natufians is strongly suggested by the numerous grinding implements and many flint blades showing silica gloss on their edges. Similar blades have been found hafted in bone handle, some of which are elaborately decorated and among the best known Natufian artifacts. The hafted flint blades that were examined, however, did not show any silica gloss. The interpretable wear to on the bone hafts is limited, as these tools were probably in use for a very ling time and suffered many random scuffs and abrasions. Several specimens, however, show increased polishing and rounding on the leading edge of the haft along the margins of the longitudinal groove into which the flint blades were set. These bone hafts are usually interpreted as sickle-hafts, but the evidence that this is their only function is not compelling. Wear of the kind observed could be most likely expected if the tool were moved longitudinally back-and-forth in the manner of a knife. It seems probable that these tools were general purpose knives.

The hunting and fishing implements are few in number. It must be recognized, however, that many are elaborate, fully worked tools which took much time to make. They would consequently have been carefully preserved. The long lifespan of some of the Ojibwa tools may be recalled in this context. As such implements would have been in use for a relatively long period of time in comparison to simpler tools they would have been considerably more important than their low numbers might suggest.

[23] In form these hooks are remarkably similar to those found in the Maglemosian (Clark 1953: 119).

Summary

It may be seen that although the specific tool forms were quite different both the Natufian and Zagros Protoneolithic peoples had developed similar crafts, and that the principal function of bone implements was in support of these crafts. It cannot be determined whether these similarities are the result of the convergence of independently conceived ideas or perhaps due to stimulus diffusion. In view of the marked differences between Natufian and Zagros Protoneolithic bone tools, both in form and means of manufacture, the direct diffusion of craft techniques does not seem likely.

Given the nature of the evidence presented, it would be foolish to attempt to give an order of importance to the various crafts and food-getting activities represented by the bone tools described here. It is too difficult to assign a specific use to an individual implement; such assignments can only be made in terms of probabilities. This difficulty, and the fact that the different activities require variable numbers of implements with differing lifespans, makes it pointless to present implement counts as an index to the prominence of the various activities. However, the best attested of these activities would be the working of leather, basketry, and matting or weaving, and the rapid appearance of large numbers of bone tools in these cultures must be associated with the new development and elaboration of these crafts. Furthermore, these tools imply a rich material culture for which most of the direct evidence has perished. Based upon the bone tool evidence, both the Natufian and the Zagros Protoneolithic, although quite different, were at much the same level of technological development. The beginnings of a settled existence were apparently accompanied by the beginning or elaboration of those domestic crafts which did much to broaden and enhance such a settled life. In terms of material culture the Natufian is the richer of the two, yet in leading a hunting and gathering existence the Natufian reveals its roots in Upper Paleolithic culture. The Zagros Protoneolithic culture shows a more clear-cut break with the past.

CHAPTER IX

The Value of Wear-pattern Studies on Bone Implements

This study has been intended to test whether, through close study of bone implements and through the microscopic study of the wear appearing on these implements, it would be possible to reach reasonable conclusions as to the purpose for which these tools were used. Given such conclusions a better idea might be gained of the level of technology of the tools' makers and the way in which that technology might be related to food-getting, crafts, and lifeways in general.

The preceding chapters show that wear-pattern studies upon bone implements can provide useful information to broaden our knowledge of the people who used these tools. Reasonably secure conclusions, however, may be reached only by weighing the evidence to be gleaned from the largest possible sample of bone implements. Few individual specimens can be assigned to a specific function with absolute certainty; conclusions must be based upon the probabilities arising from the piling up of evidence from many specimens. Through the use of common sense, comparison with objects from ethnographic sources, and particularly through comparison with experimental objects, it is often possible to describe the manner in which a tool was used. A determination of precise function is far more difficult. Within these limitations this study may be considered successful.

The bone tool assemblages studied in this project are unusual in that they document a very important stage of the technological development of society. For various historic reasons, however, very few of them have ever been studied in detail. Indeed, many important specimens have never been published even though they were excavated many years ago. Consequently a large-scale study devoted to these implements alone seemed justified. This study simultaneously provided a body of material upon which to work out and test methods for the detection and analysis of wear-patterns on bone tools.

In brief, the accomplishments of this project may be summarized as follows:

1. The project proved very successful in detecting the marks left on bone objects during the manufacture process, and in differentiating between the various manufacture processes. The majority of bone specimens exhibited interpretable surface markings left from the manufacture process. Comparisons and conclusions based on these markings may be considered the most secure of those put forward in this study.

2. Many bone objects exhibit markings from use, and where these were present the *manner* in which a tools was used (that is, the specific movements involved in that tool's use) could be determined with considerable confidence. Where scratches are present, as they are on many objects, the direction of tool movement during its use could be unambiguously determined. If we accept that the function of an entire class of tools can be determined based upon the discovery of wear on a substantial proportion of those implements, then the manner of use of nearly all the tool forms studied here could be determined with little ambiguity.

3. It proved far more difficult to distinguish the exact nature of the material manipulated with a given implement. In such cases where implements of essentially similar form, such as pointed implements, may have been used in very similar manners to work different materials, the determination of the exact use of these tools was difficult indeed. In this case secondary criteria (such as tip size) were required to reach decisions among the potential functions. Even so, these conclusions seem reasonably secure, although far less confidence can be assigned to them than in cases 1 and 2, above.

4. An attempt has been made to apply these data to permit cultural inferences about the Natufian and Zagros Protoneolithic peoples. Some of these inferences can be considered very secure. It is very clear that these bone tools were used in support of a very diversified array of craft activities. The general nature of these crafts can also be outlined fairly securely. Except in a few rare cases the precise details of the crafts for which these tools were used cannot be provided by a study such as this.

BIBLIOGRAPHY

Ahrensburg, Baruch and Ofer Bar-Yosef
1973 Human Remains from Ein Gev I, Jordan Valley, Israel. *Paleorient* 1: 201-206.

Ascher, Robert
1961 Experimental Archaeology. *American Anthropologist* 63: 793-816.

Aufenanger, Heinrich
1961 The Use of Bones (Central Highlands, New Guinea). *Anthropos* 56: 869-882.

Barker, James
1953 Memoria sobre la Cultura de los Guaika. *Boletin Indiginista Venezolano* 1: 483-489.

Barrett, S. A.
1933 Miwok Material Culture. *Bulletin of the Public Museum of Milwaukee* 2: 117-376.

Bar-Yosef, O.
1970 *The Epipaleolithic Cultures of Palestine.* Ph.D. dissertation, Hebrew University of Jerusalem.

1975 The Epipaleolithic in Palestine and Sinai. In *Problems in Prehistory: North Africa and the Levant.* F. Wendorf and A. Marks, eds. Pp. 363-378. Dallas, Texas: Southern Methodist University Press.

Bar-Yosef, O. and N. Goren
1973 Natufian Remains from Hayonim Cave. *Paleorient* 1: 49-68.

Bar-Yosef, O. and E. Tchernov
1967 Archaeological Finds and the Fossil Faunas of the Natufian and Microlithic Industries at Hayonim Cave (Western Galilee), Israel. *Israel Journal of Zoology* 15: 104-140.

1970 The Natufian Bone Industry of ha-Yonim Cave. *Israel Exploration Journal* 20: 1410150.

Basedow, H.
1925 *The Australian Aboriginal.* Adelaide: F. W. Preece and Sons.

Best, C.
1924 The Maori, Volume 2. *Memoirs of the Polynesian Society*, 5.

Biberson, P., and E. Aguirre
1965 Expériences de taille d'outils préhistoriques dans des os d'éléphant. *Quatenaria* 7: 165-183.

Billamboz, André
1977 L'Industrie du bois de cerf en Franche-Compté au Néolithique et au début de l'Age du Bronze. *Gallia Préhistoire* 20: 91-176.

Binford, L.
1981 *Bones: Ancient Men and Modern Myths.* New York: Academic Press.

Bonifay, E.
1974 Outils sur os et os utilisés dans le Paléolithique ancien du Mas des Caves à Luniel-Viel (Hérault). Note Préliminaire. In *Premier Colloque International de l'Os dans la Préhistoire.* Henriette Camps-Fabrer, ed. Pp. 157-167. Aix-en-Provence: Editions de l'Université de Provence.

Bordaz, Jacques
1969 Reflections on Typology and Technology in the Paleolithic. *Arctic Anthropology* 6: 1-29.

Bostanci, E. Y.
1959 Researches on the Mediterranean Coast of Anatolia: A New Paleolithic Site at Beldibi. *Anatolia* 4: 129-178.

1965 The Mesolithic at Beldibi and Belbasi. *Anthropoloji* 3: 91-134.

Bouchud, J.
1974 Les Traces de l'activité humaine sur les os fossiles. In *Premier Colloque International de l'Os dans la Préhistoire.* Henriette Camps-Fabrer, ed. Pp. 27-34. Aix-en-Provence: Editions de l'Université de Provence.

Braidwood, R. J., and B. Howe
1960 Investigations in Iraqi Kurdistan. *Studies in Ancient Oriental Civilizations, no. 31.* Chicago: University of Chicago Press.

Breuil, H.
1938 The Use of Bone Implements in the Old Paleolithic Period. *Antiquity* 12: 56-67.

1939 Bone and Antler Industry of the Choukoutien Sinanthropus Site. *Paleontologia Sinica*, n. s., D. 6.

Buck, P.
1952 *The Coming of the Maori.* Wellington, N. Z.: Whitcomb and Tombs, Ltd.

Campana, Douglas V.
1977 Making Stereo-Photomacrographs for Archaeological Studies. *Journal of Field Archaeology* 4: 435-440.

Camps-Fabrer, Henriette
1974 Projet de code pour l'analyze des objets en os. *Banques de Données Archéologiques. Colloques Nationaux de Centre National de la Recherche Scientifique.* Paris: CNRS.

Camps-Fabrer, Henriette, ed.
1974 *Premier Colloque International de l'Os dans la Préhistoire.* Aix-en-Provence: Editions de l'Université de Provence.

1976 *Méthodologie appliquée à l'industrie de l'os préhistorique.* Paris: CNRS.

Cauvin, J.
1972 Nouvelles fouilles à Tell Mureybet (Syrie): 1971-1972. Rapport préliminaire. *Annales Archéologiques Arabes de Syrie* 22: 105-115.

Childe, Vere Gordon
1952 *New Light on the Most Ancient East.* 4th ed. London: Routledge & Kegan Paul.

Clark, J. G. D.
1952 *Prehistoric Europe: The Economic Basis.* London: Methuen.

1953 *The Mesolithic Settlement of Northern Europe: A Study of the Food-Gathering Peoples of Northern Europe During the Early Post-Glacial Period.* New York: Greenwood Press.

1954 *Excavations at Star Carr: An Early Mesolithic Site at Seamer near Scarborough, Yorkshire.* Cambridge: Cambridge University Press.

Clark, J. G. D., and M. W. Thompson
1954 The Groove and Splinter Technique of Working Antler in Upper Palaeolithic and Mesolithic Europe, with Special Reference to the Material from Star Carr. *Proceedings of the Prehistoric Society* 19: 148-160.

Cline, W., *et al.*
1938 The Sinkaretk or Southern Okanacon of Washington. *Contributions from the Laboratory of Anthropology, General Series in Anthropology, No. 6.* L. Spier, ed. Menasha, Wisconsin: George Benton Publishing Company.

Coles, J.
1973 *Archaeology by Experiment.* London: Hutchinson University Library.

Coon, C. S.
1955 *The History of Man.* London: Jonathan Cape.

Dart, R. A.
1925 Australopithecus africanus: The Man-ape of South Africa. *Nature* 115: 195-199.

1957 *The Osteodontokeratic Culture of Australopithecus prometheus.* Memoir of the Transvaal Museum, No. 10. Pretoria, South Africa.

1965 The Unavoidable Osteodontokeratic Culture. In *Dr. D. N. Wadia Commemorative Volume.* A. G. Jhingran, ed. Pp. 231-254. Calcutta: Mining, Geological and Metallurgical Institute of India.

Dauvois, M.
1974 Industrie osseuse préhistorique et expérimentations. In *Premier Colloque International de l'Os dans la Préhistoire.* Henriette Camps-Fabrer, ed. Pp. 73-84. Aix-en-Provence: Editions de l'Université de Provence.

Densmore, F.
1928 *Chippewa Customs.* Smithsonian Institution Bureau of American Ethnology, Bulletin no. 86.

Elbashan, D.
1966 Monthly Rainfall Isomers in Israel 1931-1960. *Israel Journal of Earth Sciences* 15: 1-7.

D'Errico, F., and G. Giacobini
1985 Approche Méthodologique de L'Analyse de L'Outillage Osseux: un Example d'Étude. *L'Anthropologie.* 89 (4): 457-472.

D'Errico, F., G. Giacobini and P. F. Puech
1984 Les Réplications en Vernis des Surfaces Osseusses Façonnées: Études Expérimentales. *Bulletin de la Société Préhistorique Française* 81 (6): 169-170.

1986 Varnish Replicas: A New Method for the Study of Worked Bone Surfaces. *Ossa* 9 (11): 29-51.

Evans, F. G.
1973 Factors Affecting the Mechanical Properties of Bone. *Bulletin of the New York Academy of Medicine* 49: 751-764.

Evans, Sir John
1872 *Ancient Stone Implements, Weapons, and Ornaments of Great Britain.* New York: Appleton.

Feustel, Rudolph
1973 *Technik der Steinzeit: Archäeolithicum-Mesolithicum.* Weimar: H. Böhlau.

Garrod, D. A. E.
1930 Note on Three Objects of Mesolithic Age from a Cave in Palestine. *Man* 30: 77-78.

1932 A New Mesolithic Industry, The Natufian of Palestine. *Journal of the Royal Anthropological Institute* 42: 257-269.

1957 The Natufian Culture: The Life and Economy of a Mesolithic People in the Near East: Albert Reckett Archaeological Lecture, British Academy. *Proceedings of the British Academy* 43.

Garrod, D. A. E., and D. M. Bate
1937 *The Stone Age of Mount Carmel*, I. Oxford, Oxford University Press.

Gould, Richard A. ed.
1978 *Explorations in Ethnoarchaeology.* Albuquerque: University of New Mexico Press.

Graham, J. A., Heizer, R. F. and T. R. Hester
1972 *A Bibliography of Replicative Experiments in Archaeology.* Berkeley: Archaeological Research Facility, University of California.

Grinnel, G. B.
1901 The Lodges of the Blackfeet. *American Anthropologist* 3: 650-668.

Gusinde, M.
1931 Die Feuerland Indianer. *Anthropos* 1.

Hecker, H.
1975 *The Faunal Analysis of the Primary Food Animals from Pre-Pottery Neolithic Beidha.* Ph.D. dissertation, Columbia University.

Henry, D.
1973a *The Natufian Culture of Palestine: Its Material Culture and Ecology.* Ph.D. dissertation, Southern Methodist University.

1973b The Natufian Site of Rosh Zin: A Preliminary Report. *Palestine Exploration Quarterly* 105: 129-140.

1976 The Excavation at Hayonim Terrace: An Interim Report. *Journal of Field Archaeology* 3: 391-406.

Henry, D., and A. Frank Servello
1974 Compendium of Carbon-14 Determinations Derived from Near Eastern Prehistoric Deposits. *Paleorient* 2: 19-44.

Holley, G. A., and T. A. Del Bene
1981 An Evaluation of Keeley's Microwear Approach. *Journal of Archaeological Science* 8: 337-352.

Jenness, D.
1937 Arrow-straighteners, Thong-smoothers, and Batons-de-commandement. *Man* 37: 73-74.

1946 *Material Culture of the Copper Eskimo.* Report of the Canadian Arctic Expedition 1913-18, Volume 16. Ottawa: Edmond Cloutier.

Jochelson, W.
1925 *Archaeological Investigations in the Aleutian Islands.* Washington: Carnegie Institute.

Johnson, E.
1985 Current Developments in Bone Technology. *Advances in Archaeological Method and Theory* 8, M. Schiffer, ed. Pp. 157-233. New York: Academic Press.

Johnson, L. Lewis
1978 A History of Flint-knapping Experimentation, 1838-1976. *Current Anthropology* 19: 337-372.

Kantman, S.
1969 Experimental Import or Observation and the Significance of Functionalism in Paleolithic Typology - An Essay. *Quaternaria* 11: 263-273.

Katz, J. L.
1971 Hard Tissue as a Composite Material - I. Bounds on the Elastic Behavior. *Journal of Biomechanics* 4: 108-126.

Keeley, L. H.
1977 The Functions of Paleolithic Stone Tools. *Scientific American* 237: 108-126.

Keeley, L. H., and M. H. Newcomer
1977 Microwear Analysis of Experimental Flint Tools: A Test Case. *Journal of Archaeological Science* 4: 29-62.

Kenyon, K.
1959 Earliest Jericho. *Antiquity* 33: 5-9.

Kirkbride, D.
1966 Five Seasons at the Pre-Pottery Neolithic Village of Beidha in Jordan. *Palestine Exploration Quarterly* 98: 8-72.

Kluckhohn, C., Hill, W. W. and L. W. Kluckhohn
1971 *Navaho Material Culture.* Cambridge, Mass.: Harvard University Press.

Kramer, Carol, ed.
1979 *Ethnoarchaeology: Implications of Ethnography for Archaeology.* New York: Columbia University Press.

Kroeber, A. L.
1925 *Handbook of the Indians of California.* Smithsonian Institution Bureau of American Ethnology, Bulletin No. 78.

Legge, A. J.
1972 Prehistoric Exploitation of the Gazelle in Palestine. In *Papers in Economic Prehistory.* E. S. Higgs, ed. Pp. 19-24. Cambridge: Cambridge University Press.

Lyman, R. L.
1984 Broken Bones, Bone Expediency Tools, and Bone Pseudotools: Lessons from the Blast Zone Around Mount St. Helens, Washington. *American Antiquity* 49 (2): 315-333.

Lynch, B. D., and T. F. Lynch
1968 The Beginnings of a Scientific Approach to Prehistoric Archaeology in 17th and 18th Century Britain. *Southwestern Journal of Anthropology* 24: 33-65.

MacGregor, Arthur M.
1985 *Bone, Antler, Ivory and Horn: The Technology of Skeletal Materials Since the Roman Period.* Totowa, New Jersey: Barnes and Noble.

Mason, Otis T.
1904 *Aboriginal American Basketry: Studies in a Textile Art Without Machinery.* Washington D. C.: Government Printing Office.

Massoulard, E.
1949 *Préhistoire et Protohistoire d'Egypte.* Paris: Institute d'Ethnologie.

McKenna, P. J.
1984 Faunal Remains. The Architecture and Material Culture of Chaco Canyon, New Mexico. *Reports of the Chaco Center*, 7: 313-345. National Park Service.

Mellaart, J.
1967 *Çatal Hüyük, a Neolithic Town in Anatolia.* New York: McGraw-Hill.

1970 *Excavations at Haçilar.* Edinburgh: University Press.

1975 *The Neolithic of the Near East.* New York: Charles Scribner's Sons.

Moore, A. M T.
1975 The Excavation of Tell Abu Hureyra in Syria: A Preliminary Report. *Proceedings of the Prehistoric Society* 41: 50-77.

Morrow, Mable
1975 *Indian Rawhide: An American folk Art.* Norman: University of Oklahoma Press.

Myers, T. P., M. R. Voorhies, and R. G. Corner
1980 Spiral Fractures and Bone Pseudotools at Paleontological Sites. *American Antiquity* 45 (3): 483-490.

Neuville, R.
1951 *Le Paléolithique et Mésolithique de Désert de Judée.* Archives de l'Institut de la Paléontologie Humaine (Paris), Memoire 24.

Newcomer, M. H.
1974a Outils en Os de Paléolithique Supérieur de Ksar Akil (Liban). In *Premier Colloque International de l'Os dans la Préhistoire.* Henriette Camps-Fabrer, ed. Pp. 59-65. Aix-en-Provence: Editions de l'Université de Provence.

1974b Study and Replication of Bone Tools from Ksar Akil (Lebanon). *World Archaeology* 6: 38-153.

Noy, T., Higgs, E. Legge, A. and I. Gisis
1973 Recent Excavations at Nahal Oren, Israel. *Proceedings of the Prehistoric Society* 39: 75-99.

Odell, G. H.
1975 Micro-wear in Perspective: A Sympathetic Response to Lawrence H. Keeley. *World Archaeology* 7: 226-240.

Peltier, A.

1986 Étude Expérimentale des Surfaces Osseuses Façonnées et Utilisées. *Bulletin de la Société Préhistorique Française* 1-7.

Perkins, D.
1964 Prehistoric Fauna from Shanidar, Iraq. *Science* 144: 1565-1566.

Perrot, J.
1966 Le Gisement Natoufien de Mallaha (Eynan), Israel. *L'Anthropologie* 70: 437-484.

1968 La Préhistoire Palestinienne. In *Supplément au Dictionaire de la Bible*, Vol. 8, 43. Pp. 286-446. Paris: Letouzey et Ane.

Pope, M. H., and J. O. Outwater
1972 The Fracture Characteristics of Bone Substance. *Journal of Biomechanics* 5: 457-466.

1974 Mechanical Properties of Bone as a Function of Position and Orientation. *Journal of Biomechanics* 7: 61-66.

Poplin, F.
1974 Principes de la Détermination des Matières Dures Animales. In *Premier Colloque International de l'Os dans la Préhistoire*. Henriette Camps-Fabrer, ed. Pp. 15-20. Aix-en-Provence: Editions de l'Université de Provence.

Pradel, L.
1972 Nomenclature et Possibilités Functionelles de l'Outillage en Pierre du Paléolithique en France. *Quartär* 23: 37-51.

Radcliffe-Brown, A. R.
1922 *The Andaman Islanders: A Study in Social Anthropology*. Cambridge: Cambridge University Press.

Ray, V. F.
1933 *The Sanpoil and Nespelem: Salishin Peoples of Northeastern Washington*. University of Washington Publications, Vol. 5. Seattle: University of Washington Press.

Rigaud, A.
1972 La Technologie du Burin Appliquée au Materièl Osseux de la Garenne (Indre). *Bulletin de la Préhistorique Française* 69: 104-108.

Rincon, A. and E. Aguirre
1974 Analyse Comparative et Différentielle des Assemblages de Fragments Osseux pour Vérification d'Hypothèses d'Action Culturelle. In *Premier Colloque International de l'Os dans la Préhistoire*. Henriette Camps-Fabrer, ed. Pp. 111-117. Aix-en-Provence: Editions de l'Université de Provence.

Rodden, R. J.
1962 Excavations at the Early Neolithic Site at Nea Nikomedia, Greek Macedonia (1961 Season). *Proceedings of the Prehistoric Society* 28: 267-288.

Sadek-Kooros, H.
1972 Primitive Bone Fracturing: A Method of Research. *American Antiquity* 37: 369-381.

Semenov, S. A.
1964 *Prehistoric Technology*. English Edition. London: Cory, Adams, and Mackay.

1970 The Form and Function of the Oldest Stone Tools (A Reply to Prof. Bordes). *Quartär* 21: 1-20.

1973 Fonctionologie du Paléolithique. In *Actes du VIIIême Congrès International des Sciences Préhistoriques et Protohistoriques*, II. Pp. 108-113. Beograd.

Shipman, P., G. Foster, and M. Schoeninger
1984 Burnt Bones and Teeth: an Experimental Study of Color, Morphology, Crystal Structure and Shrinkage. *Journal of Archaeological Science* 11: 307-325.

Sisson, S. and J. D. Grossman
1975 *The Anatomy of the Domestic Animals*. 5th ed. Robert Getty, ed. Philadelphia: W. B. Saunders.

Solecki, Ralph S.
1964 Shanidar Cave, A Late Pleistocene Site in Northern Iraq. In *Report of the VIth International Congress on the Quaternary*, Vol. 4. Warsaw.

Solecki, Rose L.
1981 *An Early Village Site at Zawi Chemi Shanidar*. Bibliotheca Mesopotamica 13. Malibu: Undena Publications.

1964 Zawi Chemi Shanidar, A Post-Pleistocene Village Site in Northern Iraq. In *Report of the VIth International Congress on the Quaternary*, Vol. 4. Warsaw.

Solecki, R. S. and R. L. Solecki
1963 Two Bone Hafts from Northern Iraq. *Antiquity* 37: 58-60.

Solecki, R. L. and R. S. Solecki
1970 Grooved Stones from Zawi Chemi Shanidar, A Protoneolithic Site in Northern Iraq. *American Anthropologist* 72: 831-841.

Steinbring, J.
1966 The Manufacture and Use of Bone Defleshing Tools. *American Antiquity* 31: 575-581.

Stekelis, M., and O. Bar-Yosef
1965 Un Habitat du Paléolithique Supérieur a Ein Guev (Israel), Note Préliminaire. *L'Anthropologie* 69: 176-183.

Stekelis, M. and T. Yizraeli
1963 Excavations at Nahal Oren (Fallah), Preliminary Report. *Israel Exploration Journal* 13: 1-12.

Stordeur, D., Anderson-Gerfaud, P., Deraprahamian, G., Reynolds, P., and L. Le Ribault
in press Les Omoplates encochées néolithiques de Ganj Dareh (Iran):étude morphologique et foncionelle. *Cahiers de l'Euphrate* 4.

Stordeur, D.
1981 L'outil d'os pendant le préhistoire. *La Recherche* 121: 452-465.

1980 La Contribution de l'Industrie de l'Os à Délimitation des Aires Culturelles: L'Exemple du Natoufien. *Préhistoire du Levant: Chronologie et Organization de L'Espace depuis les Origines jusque'au VIᵉ Millénaire.* Lyon: CNRS.

1978 Proposition de classement des objets en os selon le degrée de transformation imposé à la matière première. *Bulletin de la Société Préhistorique Française* 75 (1): 20-23.

Stordeur-Yedid, D.
1979 *Les Aiguilles à Chas au Paléolithique.* XIII Supplement to *Gallia Préhistoire.* Paris: CNRS.

Thomas, E. M.
1959 *The Harmless People.* New York: Alfred A. Knopf.

Thompson, M. W.
1964 Translator's Preface. In *Prehistoric Technology.* English Edition. S. A. Semenov. London: Cory, Adams, and Mackay.

Tringham, R. E., Cooper G., Odell G., Voytek, B., and A. Whitman
1974 Experimentation in the Formation of Edge Damage. *Journal of Field Archaeology* 1: 171-196.

Turville-Petre, F.
1932 Excavations at the Mugharet El-Kebarah. *Journal of the Royal Anthropological Institute* 62: 270-276.

Tyzzer, E. E.
1936 The "Simple Bone Point" of the Shell-heaps of the Northeastern Algonkian Area and Its Probable Significance. *American Antiquity* 4: 261-279.

Unger-Hamilton, R.
1989 The Epi-Palaeolithic Southern Levant and the Origins of Cultivation. *Current Anthropology* 30 (1): 88-103.

Van Zeist, W.
1969 Reflections on Prehistoric Environments in the Near East. In *The Domestication and Exploitation of Plants and Animals.* Peter J. Ucko and G. W. Dimbleby, eds. Pp. 35-46. Chicago: Aldine.

Vaughan, Patrick C.
1985 *Use-Wear Analysis of Flaked Stone Tools.* Tucson: University of Arizona Press.

Walker, Philip L. and Jeffrey C. Long
1977 An Experimental Study of the Morphological Characteristics of Tool Marks. *American Antiquity* 47 (4): 605-616.

Wallace, W. D., and R. S. Wallace
1955 *The Micmac Indians of Eastern Canada.* Minneapolis: University of Minnesota Press.

Wissler, Clark
1910 *Material Culture of the Blackfoot Indians.* Anthropological Papers of the American Museum of Natural History, Vol. 5. Part 1.

Wreschner, E., Avnimelech, M., Schmid, E., Haas, G., and R. A. Dart
1967 The Geula Caves, Mount Carmel. *Quarternaria* 9: 69-140.

Yellen, J.
1977 *Archaeological Approaches to the Present: Models for Reconstructing the Past.* New York: Academic Press.

Yesner, D. R. and R. Bonnichsen
1979 Caribou Metapodial Shaft Splinter Technology. *Journal of Archaeological Science* 6: 303-308

Zohary, D.
1969 The Progenitors of Wild Wheat and Barley in Relation to Domestication and Agricultural Dispersal in the Old World. In *The Domestication and Exploitation of Plants and Animals.* Peter J. Ucko and G. W. Dimbleby, eds. Pp. 47-66. Chicago: Aldine.

APPENDIX

Artifact Numbers Used in This Study and Corresponding Excavator's and Museum Numbers

HAY1	H44	HAY61	H220
HAY2	H45	HAY62	H94
HAY3	H46	HAY63	H98
HAY4	H47	HAY64	H99
HAY6	H50	HAY65	H100
HAY7a, 7b	H51	HAY66	H124
HAY9	H192	HAY67	H126
HAY10	H52	HAY68	H70
HAY11	H120	HAY69	H81
HAY12	H125	HAY70	H73
HAY13	H127	HAY71	H71
HAY14	H128	HAY72	H103
HAY15	H129	HAY73	H104
HAY16	H131	HAY74	H105
HAY17	H136	HAY75	H82
HAY18	H137	HAY76	H67
HAY19	H138	HAY77	H221
HAY20	H141	HAY78	H229
HAY21	H243	HAY79	H132
HAY22	H254	HAY80	H69
HAY23	H92	HAY81	H62
HAY24	H91	HAY82	H240
HAY25	H112	HAY83	---
HAY26	H111	HAY84	H172
HAY27	H118	HAY85	H245
HAY28	H122	HAY86	H102
HAY29	H121	HAY87	H48
HAY30	H123	HAY88	H287
HAY31	H116	HAY89	H178
HAY32	H117	HAY90	H61
HAY33	H79	HAY91	H59
HAY34	H74	HAY92	H255
HAY35	H232	HAY93	---
HAY36	H281	HAY94	H177
HAY38	H284	HAY95	H182
HAY39	H285	HAY96	H183
HAY40	H286	HAY97	H187
HAY41	H288	HAY99	---
HAY42	H291	HAY100	H158
HAY43	H161	HAY101	H188
HAY44	H238	HAY102	H190
HAY47	H175	HAY103	H189
HAY48	H60	HAY104	H101
HAY49	H174	HAY105	H180
HAY50	H173	HAY107	H184
HAY51	H38	HAY108	H64
HAY52	H39	HAY109	H80
HAY53	H40	HAY110	H63
HAY54	H42	HAY111	H185
HAY55	H259	HAY112	H164
HAY57	H88	HAY113	H139
HAY58	H93	HAY114	H228
HAY60	H75	HAY115	H162
HAY116	H222	HAY148	H320
HAY117	H181	HAY149	H324
HAY118	H154	HAY150	H336

HAY119	H225
HAY120	H376
HAY121	H373
HAY122	H375
HAY123	H326
HAY124	H311
HAY125	H303
HAY126	H309
HAY127	H344
HAY128	H345
HAY129	H343
HAY130	H339
HAY131	H362
HAY132	---
HAY133	---
HAY134	---
HAY135	H319
HAY136	H37
HAY137	H307
HAY138	H315
HAY139	H348
HAY140	H369
HAY142	H293
HAY143	H294
HAY144	H301
HAY145	H304
HAY146	H316
HAY147	H318
HAY151	H341
HAY152	---
HAY153	H383
HAY154	H314
HAY155	H90
HAY156	H86
HAY157	H89
HAY158	H65
HAY159	H77
HAY160	H78
HAY161	H333
HAY179	H325
HAY180	H295
HAY191	H380
HAY192	H349
HAY193	H151
HAY194	H385
HAY195	---
HAY196	---
HAY197	H83
HAY198	H95
HAY199	H119
HAY200	H97
HAY201	H113
HAY202	H114
HAY203	H115
HAY204	H119

All Hayonim Cave specimens are located in the Hebrew University.

KEB1	33-58
KEB2	33-59
KEB3	33-60
KEB4	33-68
KEB5	33-91
KEB6	33-55
KEB7	33-92
KEB8	33-94
KEB9	33-95
KEB10	33-129
KEB11	I10704
KEB12	33-114
KEB13	33-116
KEB14	33-118
KEB15	33-120
KEB16	33-104
KEB17	33-105
KEB18	33-106
KEB19	33-111
KEB20	33-113
KEB21	33-115
KEB22	33-51
KEB23	33-54
KEB24	33-52
KEB25	33-53
KEB26	33-86
KEB27	33-87
KEB28	33-85
KEB29	33-77
KEB30	33-76
KEB31	33-73
KEB32	33-137
KEB33	33-138
KEB34	I10702
KEB35	33-45
KEB36	33-47
KEB37	33-48
KEB38	33-50
KEB39	33-49
KEB40	I10700
KEB41	I10701
KEB42	33-125
KEB43	33-103

The Kebarah Cave specimens listed above are located in the Rockefeller Museum.

KEB44	---
KEB45	---
KEB46	---
KEB47	---
KEB48	---
KEB49	---
KEB50	---
KEB51	---
KEB52	---
KEB53	---
KEB54	---
KEB55	---
KEB56	---
KEB57	---
KEB58	---
KEB67	---
KEB68	---
KEB69	---
KEB70	---
KEB71	---
KEB72	2-8-10
KEB73	2-9-11
KEB74	2-8-9
KEB75	2-8-12
KEB76	2-8-15
KEB77	2-8-16
KEB78	---
KEB79	2-8-7
KEB80	2-3-1
KEB81	2-8

KEB59	---	KEB82	2-8-3
KEB60	---	KEB83	---
KEB61	---	KEB84	2-8-6
KEB62	---	KEB85	2-8-14
KEB63	---	KEB86	2-8-13
KEB64	---	KEB87	---
KEB65	---	KEB88	---
KEB66	---	KEB89	2-8-8

The above specimens from Kebarah Cave are located in the British Museum.

SHK1	1.2233	SHK9	1.2229
SHK2	1.2232	SHK10	1.2587
SHK3	1.2236	SHK11	1.2223
SHK4	1.2234	SHK12	1.2224
SHK5	1.2227	SHK13	1.2230
SHK6	1.2226	SHK14	1.2235
SHK7	1.2238	SHK15	1.2231
SHK8	1.2237		

The above specimens from Shukbah are located in the Rockefeller Museum.

SHK16	3-6-3	SHK17	3-6-2
SHK18	3-6-5	SHK19	3-6-4
SHK20	3-6-6	SHK21	3-6-1

The above specimens from Shukbah are located in the British Museum.

WAD1	I.4675	WAD9	I.1726
WAD2	I.4689	WAD10	I.4634
WAD3	I.4706	WAD11	I.4682
WAD4	I.4707	WAD12	I.4686
WAD5	I.1719	WAD13	I.4718
WAD6	I.1722	WAD14	I.4681
WAD7	I.1724	WAD15	I.4702
WAD8	I.4135	WAD16	I.1721
WAD17	I.1792	WAD24	I.4725
WAD18	I.8798	WAD25	I.4726
WAD19	I.4692	WAD26	I.4719
WAD20	I.4727	WAD27	I.4700
WAD21	I.10721	WAD28	I.4701
WAD22	I.4703	WAD29	I.10722
WAD23	I.4724	WAD30	I.1728

The above specimens from El Wad are located in the Rockefeller Museum.

ORN1	F60285	ORN6	---
ORN2	F57/82	ORN7	F57697
ORN3	---	ORN8	F60284
ORN4	---	ORN9	---
ORN5	F49184	ORN10	---

The above specimens from Nahal Oren are located in the Israel Museum.

RAK1	R70/12	RAK3	R71567
RAK2	R71/59		

The above specimens from Rakefet are located in the Israel Museum.

ZCS1	58 IV	ZCS35	109 IV
ZCS2	54 IV	ZCS36	51 IV
ZCS3	54 IV	ZCS37	72 IV
ZCS4	54 IV	ZCS38	137 IV
ZCS5	70 IV	ZCS39	70 IV
ZCS6	---	ZCS40	65 IV
ZCS7	72 IV	ZCS41	72 IV
ZCS8	62 IV	ZCS42	58 IV
ZCS9	34 IV	ZCS43	54 IV
ZCS10	19 IV	ZCS44	72 IV
ZCS11	58 IV	ZCS45	53 IV

ZCS12	72 IV	ZCS46	45 IV
ZCS13	62 IV	ZCS47	63 IV
ZCS14	75 IV	ZCS48	36 IV
ZCS15	82 IV	ZCS50	---
ZCS16	72 IV	ZCS51	54 IV
ZCS17	---	ZCS52	58 IV
ZCS18	---	ZCS53	61 IV
ZCS19	65 IV	ZCS54	80 IV
ZCS20	77 IV	ZCS55	75 IV
ZCS21	49 IV	ZCS56	62 IV
ZCS22	80 IV	ZCS57	77 IV
ZCS23	82 IV	ZCS58	80 IV
ZCS24	---	ZCS59	187
ZCS25	127 IV	ZCS60	45/58 IV
ZCS26	70 IV	ZCS61	58 IV
ZCS27	51 IV	ZCS62	98 IV
ZCS28	88 IV	ZCS63	---
ZCS29	91 IV	ZCS64	80 IV
ZCS30	77 IV	ZCS65	51 IV
ZCS31	80 IV	ZCS66	63 IV
ZCS32	72 IV	ZCS67	58 IV
ZCS33	54 IV	ZCS68	45 IV
ZCS34	51 IV	ZCS69	53 IV
ZCS70	63 IV	ZCS83	62 IV
ZCS71	109 IV	ZCS84	54 IV
ZCS73	54 IV	ZCS85	88 IV
ZCS74	80 IV	ZCS86	82 IV
ZCS75	53 IV	ZCS87	34 IV
ZCS76	88 IV	ZCS88	79 IV
ZCS77	63 IV	ZCS89	84 IV
ZCS79	61 IV	ZCS90	58/61 IV
ZCS81	135 IV	ZCS91	61 IV
ZCS82	54 IV	ZCS92	63 IV

The above specimens from Zawi Chemi Shanidar are located at Columbia University.

ZCS93	66 IV	ZCS138	64 IV
ZCS94	66 IV	ZCS139	115 IV
ZCS96	55 IV	ZCS140	56 IV
ZCS97	28 IV	ZCS141	56 IV
ZCS98	28 IV	ZCS142	56 IV
ZCS100	77 IV	ZCS143	56 IV
ZCS101	77 IV	ZCS144	85 IV
ZCS102	77 IV	ZCS145	85 IV
ZCS103	77 IV	ZCS146	95 IV
ZCS104	77 IV	ZCS147	69 IV
ZCS105	111 IV	ZCS148	50 IV
ZCS106	57 IV	ZCS149	50 IV
ZCS107	94 IV	ZCS151	50 IV
ZCS108	69 IV	ZCS152	50 IV
ZCS109	69 IV	ZCS153	50 IV
ZCS110	69 IV	ZCS154	50 IV
ZCS111	---	ZCS155	50 IV
ZCS112	50 IV	ZCS156	74 IV
ZCS113	50 IV	ZCS157	76 IV
ZCS114	50 IV	ZCS158	76 IV
ZCS115	56 IV	ZCS159	55 IV
ZCS116	75 IV	ZCS160	55 IV
ZCS117	75 IV	ZCS161	35 IV
ZCS118	75 IV	ZCS162	47 IV
ZCS119	75 IV	ZCS163	47 IV
ZCS120	75 IV	ZCS164	47 IV
ZCS121	75 IV	ZCS165	34 III
ZCS122	79 IV	ZCS166	46 III
ZCS123	79 IV	ZCS167	31 III
ZCS124	79 IV	ZCS168	34 III
ZCS125	35 IV	ZCS169	31 III
ZCS126	80 IV	ZCS170	34 III
ZCS127	80 IV	ZCS171	35 III

ZCS128	71 IV	ZCS172	34 III
ZCS130	46 IV	ZCS173	30 IV
ZCS131	46 IV	ZCS174	60 IV
ZCS132	46 IV	ZCS175	475 III
ZCS133	81 IV	ZCS176	458 III
ZCS134	74 IV	ZCS177	457 III
ZCS135	74 IV	ZCS178	448 III
ZCS136	74 IV	ZCS179	470 III
ZCS137	90 IV		

The above specimens from Zawi Chemi Shanidar are located in the Iraqi Museum.

SHN1	181 IV	SHN5	328 IV
SHN2	109 IV	SHN6	328 IV
SHN3	107 IV	SHN8	328 IV
SHN4	259 IV	SHN11	342 IV

The above specimens from Shanidar Cave are located at Columbia University.

SHN12	139 III	SHN26	549
SHN13	138 III	SHN27	271
SHN15	131 II	SHN28	---
SHN17/18	130 III	SHN29	275
SHN19	389 IV	SHN30	179 III
SHN20	268 IV	SHN31	183 III
SHN21	172 IV	SHN32	183 III
SHN22	384 IV	SHN33	140
SHN23	---	SHN34	191
SHN24	165	SHN36	87
SHN25	169	SHN37	230 III

The above specimens from Shanidar Cave are located in the Iraqi Museum.

www.ingramcontent.com/pod-product-compliance
Lightning Source LLC
LaVergne TN
LVHW071213210926
848482LV00015B/804